The Ins and Outs of Azure VMware Solution

Deploy, configure, and manage an Azure VMware Solution environment

Dr. Kevin Jellow, D.H.L (h.c)

BIRMINGHAM—MUMBAI

The Ins and Outs of Azure VMware Solution

Group Product Manager: Rahul Nair
Publishing Product Manager: Niranjan Naikwadi
Senior Editor: Athikho Sapuni Rishana
Technical Editor: Nithik Cheruvakodan
Copy Editor: Safis Editing
Project Coordinator: Ashwin Kharwa
Proofreader: Safis Editing
Indexer: Hemangini Bari
Production Designer: Ponraj Dhandapani
Marketing Coordinator: Gaurav C
Senior Marketing Coordinator: Nimisha Dua

First published: Jan 2023

Production reference: 1091222

Published by Packt Publishing Ltd.
Livery Place
35 Livery Street
Birmingham
B3 2PB, UK.

ISBN 978-1-80181-431-7

www.packt.com

To my mother, Zelpha, and to the memory of my father, Claude Ronald Jellow, for their sacrifices and exemplifying the power of determination.

To my sons Chris, Bradley, and Gavin, who keep me motivated every day.

I also want to thank all my friends and family who have encouraged me daily.

Special thank you to all my teammates from the AVS GBB team at Microsoft and the AVS team at VMware. You guys are just awesome.

– Dr. Kevin Jellow, D.H.L (h.c)

Contributors

About the author

Dr. Kevin Jellow, D.H.L (h.c) is a result-driven Microsoft architect and solutions specialist with over 20 years of progressive career growth. He focuses on helping customers with their cloud transformation journey to Microsoft Azure. He has worked at Microsoft for the past 11 years in different roles within the Azure cloud business.

He is the father of three sons: Chris, Bradley, and Gavin, and grandfather of Cal-El. He is from the sunny island of Jamaica and is now based in Memphis, TN, and the USVI.

When he is not busy helping customers with their cloud transformation, he is off mentoring at his alma mater, Jose Marti Technical High School (Jamaica), through the non-profit **Jose Marti Alumni Association New York (JMAANY)**, where he is currently the sitting president.

About the reviewer

James Mendez is a Cloud Solutions Architect at Microsoft working with customers using Active Directory, Windows Server platforms, and various Azure cloud services. He has worked in the IT industry for 26 years obtaining various IT certifications (Microsoft, Cisco, VMware) and also holding various roles such as Sr. Systems Engineer and Lead IT Systems Architect. He has gained experiences and exposure to a variety of technologies over the years which include Scripting, Web Development, ETL Data Integration, Networking, Virtualization, and Hyper-converged infrastructure. Outside of work, he has several interests including being a musician (composing), traveling, cycling, running, reading, continuous learning, and spending time with family.

I'd like to thank:

My brother for the encouragement and support, being a mentor, and also sharing his invaluable experiences and insight throughout the introductory years of my career.

My parents for instilling a strong and honest work ethic in me and always encouraging me to invest 200% into anything I am passionate about.

The few but genuine friends I have for truly being there and believing in me.

Table of Contents

Preface   xv

Part 1: Getting Started with Azure VMware Solution (AVS)

1

Introduction to Azure VMware Solution   3

Network connectivity to AVS	4	Network and connectivity topologies	10
AVS hosts, clusters, and private clouds	5	Identity and access management	16
AVS high-level architecture	6	Business continuity and disaster recovery	16
Use cases for AVS in an enterprise	7	Design considerations for AVS BC	17
Data center footprint deduction, consolidation, and retirement	7	Design considerations for AVS DR	17
Data center expansion based on demand	8	Security, governance, and compliance	18
Disaster recovery and business continuity	8	Security	18
Speed and simplification of migration/hybrid cloud	8	Governance	19
AVS is very cost-effective	8	Compliance	19
Enterprise-scale for AVS	8	Management and monitoring	20
Prerequisites for the implementation of the enterprise-scale landing zone for AVS	9	Summary	21

2

Enterprise-Scale for AVS   23

Network and connectivity topology for AVS   23

Understanding networking requirements for AVS   25

Networking scenarios for AVS with traffic inspection   26

Identity and access management for AVS   33

vCenter privileges   34

Business continuity and disaster recovery   37

Design considerations for business continuity in AVS   37

The recommended design considerations for business continuity in AVS   38

Design considerations for AVS disaster recovery   39

Summary   41

Part 2: Planning and Deploying AVS

3

Planning for an Azure VMware Solution Deployment   45

Subscription identification   45
Resource group identification   47
Azure region   47
Region pairs in Azure   47

AVS resource name   48
Host size   48
Determining the number of hosts and clusters   48
Host quota request for AVS   49
Requesting a /22 CIDR IP segment for AVS management components   52

Defining the AVS workload network segments   54
Defining the virtual network gateway   55

VMware HCX   56
Why use VMware HCX?   56
Defining VMware HCX network segments   56

HCX appliance IP requirements   57
HCX port requirements   57
Benefits of extending the L2 network to AVS   58

Summary   60

4

Deploying an Azure VMware Solution Cluster   63

Prerequisites to deploy AVS   63
Registering the AVS resource provider   64

Deploying an AVS cluster   66
The basic information needed to deploy AVS   66

AVS deployment validation   69

Connecting AVS to your Azure infrastructure   70

Validating the connection between AVS and Azure   73

Connecting AVS to your on-premises environment   76

Creating an ExpressRoute authorization key on your on-premises ExpressRoute circuit   76

Peering AVS with your on-premises environment   77

Validating the connection between AVS and on-premises   78

Summary   79

5

Deploying and Configuring HCX in Azure VMware Solution   81

Prerequisites for deploying HCX Advanced   82

Deploying HCX Advanced using the Azure portal   82

Downloading and deploying the VMware HCX Connector OVA   84

Prerequisites   84

Downloading the HCX Connector OVA file   84

Deploying the VMware HCX Connector OVA   86

Activating VMware HCX   87

Configuring the on-premises HCX Connector   90

Adding a site pairing   90

Creating network profiles   91

Creating a compute profile   93

Creating a Service Mesh   98

Summary   103

6

Networking in AVS using NSX-T   105

Configuring DHCP for AVS   107

Prerequisites   107

Using the Azure portal to create a DHCP server or relay for AVS   107

Adding an NSX-T segment using the Azure portal   112

Adding an NSX-T segment using NSX-T Manager   114

Verifying the newly created network segment   116

Configuring DNS for AVS   117

Configuring a DNS forwarder   118

DNS name resolution verification for AVS   120

Deploying a test VM and connecting it to the newly created segment   122

Moving a VM to a different network segment   125

Summary   128

Part 3: Configuring Your AVS

7

Creating and Configuring a Secure vWAN Hub for Internet Connectivity   131

Azure vWAN in Azure	131	Deploying Azure Firewall to secure the vWAN	142
Advantages of a vWAN	133	Prerequisites	142
vWAN types	133	Deploying Azure Firewall	143
Creating a vWAN in Azure	134	Creating an Azure Firewall policy for the AVS internet connection	146
Prerequisites	134	Adding a rule to the firewall policy	147
Creating a virtual hub and an ExpressRoute Gateway in the vWAN	136	Associating the Firewall policy with the hub	148
Connecting the AVS ExpressRoute circuit to the hub gateway	139	Routing the AVS traffic to the vWAN hub	149
Changing the size of the gateway	141	Summary	152

8

Inspecting Traffic for AVS   153

Internet consideration design options for AVS	153	Deploying an Azure Route Server	162
		Deploying a Quagga using an NVA	163
SNAT managed from AVS	154		
Public IP to the AVS NSX Edge	154	Configuring the Route Server peering	168
Some considerations for which option you choose to utilize	155	Checking the learned routes on the Route Server and the Quagga NVA	170
Implementing an NVA solution for traffic inspection	156	Summary	172
Prerequisites	157		
Creating a virtual network	157		

9

Storage Concepts in AVS   173

Fault tolerance and storage policies   174
Configuring a storage policy   174
Prerequisites   175
Listing all storage policies   175
Setting a storage policy for a VM   178
Specifying the default storage policy for an
AVS cluster   181
Encryption of data at rest   184

Azure NetApp Files   185
Prerequisites   185
Creating a NetApp Files volume for AVS   185

Creating a NetApp account   186
Creating a capacity pool for Azure NetApp
Files   188
Delegating a subnet to Azure NetApp Files 190
Creating an NFS volume for Azure NetApp
Files   190
Attaching an Azure NetApp Files volume to
an AVS cluster   194
Supported regions   195
Performance best practices   195

Summary   198

10

Working with VMware Site Recovery Manager   199

Understanding what SRM in AVS is   200
Business continuity and disaster recovery 200
Identifying your company's business
continuity and disaster recovery needs   200
Supported scenarios for SRM   201
BCDR support types used by SRM   202

Installing SRM in your primary and
secondary AVS environments   203
Deploying SRM in AVS   203
Installing the vSphere Replication appliance 204

Configuring site pairing for vCenter   206
Configuring site pairing in vCenter   206

Connecting the SRM instances on
both the protected and recovery sites   210
Configuring mapping between both the
primary and secondary SRM sites   210
Creating a new network mapping   212
Configuring virtual machine replication   216
Creating and managing protection groups for
SRM   220
Testing and running a recovery plan   223
Running a recovery plan   224

Summary   225

Part 4: Governance and Management for AVS

11

Managing an Azure VMware Solution Environment   229

AVS business alignment	231	VMware Syslogs configuration for AVS	244
Managing and monitoring your AVS environment	231	Prerequisites	244
		Diagnostic settings configuration	245
Configuring Azure Alerts for AVS	235	Summary	247

12

Leveraging Governance for Azure VMware Solution   249

A unified security and compliance approach	250	AVS environment governance	258
Integrating Azure-native tools/ services with AVS	251	Governance for workload applications and VMs	259
Security for your AVS environment	252	Azure-native solutions integration	260
Security for identity	252	Use cases for Azure Files	261
Network security	253	Key advantages	261
VM and guest application security	255	How to create an Azure file share	262
Compliance	256	Prerequisites	263
		Creating an Azure file share	263
Governance	258	Mapping the Azure file share to an AVS VM	266
		Summary	268

13

Summary of Azure VMware Solution, Roadmap, and Best Practices   269

AVS overview	270	Network and connectivity topology for AVS	274
AVS hardware and software specification	270		
AVS high-level architecture	272	Understanding networking requirements for AVS	276
Use cases for AVS in an enterprise	273		

Networking scenarios for AVS with traffic inspection   277
Planning for an AVS Deployment   279
Subscription identification   279
Resource group identification   280
Azure region   281
Region pairs in Azure   281
AVS resource name   281
Determining the number of nodes   281
Host quota request for AVS   282

Requesting a /22 address space for AVS management components   282
Defining the AVS workload network segments   283
Defining the virtual network gateway   284
Managing an AVS environment   285
Leveraging governance for AVS   286

AVS roadmap   287
Best practices for planning, deploying, and managing AVS   288
Architectural design for AVS   289

Summary   291

Index   293

Other Books You May Enjoy   302

Preface

Azure VMware Solution (**AVS**) is a first-party Microsoft Azure service developed in conjunction with VMware that provides a familiar vSphere-based, single-tenant private cloud on Azure that is like the one used by VMware. The VMware technology stack consists of the following components: vSphere, NSX-T, vSAN, and HCX. AVS is installed on dedicated infrastructure in Azure data centers and runs natively on that infrastructure. In comparison to existing on-premises VMware infrastructures, AVS provides a consistent and well-known user experience. Customers may deploy an AVS environment in a matter of hours and migrate **virtual machine** (**VM**) resources in a matter of minutes. Microsoft supplies all the networking, storage, management, and support services that are required.

By the end of this book, you will have learned how to plan, deploy, and configure an AVS environment for real-world results.

Who this book is for

This book is intended for VMware administrators, cloud solutions architects, and anyone interested in learning how to deploy, configure, and manage an AVS environment in Azure. It is also for technology leaders who want to get out of the data center business or expand their on-premises data center into Microsoft Azure.

This book's readers should already be familiar with VMware solutions and understand Azure networking.

What this book covers

Chapter 1, *Introduction to Azure VMware Solution*, explains how AVS provides a consistent, well-known user experience with existing on-premises VMware environments. Customers can deploy an AVS environment in just a few hours and quickly migrate VM resources.

Chapter 2, *Enterprise-Scale for Azure VMware Solution*, is all about the open source Azure Resource Manager and Bicep templates in the Enterprise-scale scenario for AVS. The Enterprise-scale implementation follows the architecture and best practices of the Cloud Adoption Framework's Azure landing zones, focusing on enterprise-scale design concepts.

Chapter 3, Planning for an Azure VMware Solution Deployment, identifies and acquires everything that you need for your deployment throughout the planning stage since for a successful production-ready environment for building VMs and migration, planning your AVS deployment is crucial.

Chapter 4, Deploying Your First Azure VMware Solution Cluster, assists you with learning about AVS ideas, identifying AVS prerequisites, planning for the initial deployment, creating the first AVS private cloud, and connecting an on-premises data center to the AVS **software-defined data center (SDDC)**.

Chapter 5, Deploying and Configuring HCX in Azure VMware Solution, teaches you how to deploy and configure HCX Advanced in your on-premises vCenter.

Chapter 6, Adding Network Segments in Azure VMware Solution, guides you on how to configure NSX-T network segments using NSX-T Manager or the Azure portal after a successful AVS private cloud deployment. The segments are logical switches that your AVS workloads require.

Chapter 7, Creating and Configuring a Secure vWAN Hub for Internet Connectivity, discusses how to connect to the internet via a Virtual WAN, given that utilizing VMware's SDDC in conjunction with the Azure cloud ecosystem necessitates a distinct set of architectural considerations for cloud-native and hybrid situations.

Chapter 8, Inspecting Traffic for AVS, details how, when migrating to AVS, customers may want to preserve operational continuity with their existing third-party networking and security solutions. The communication mechanism has nothing to do with the NSX-T service insertion/network introspection certification process for vSphere or AVS, and third-party platforms may include products from Cisco, Juniper, or Palo Alto Networks.

Chapter 9, Adding Additional Storage to the AVS Datastore, walks you through the process of deploying an Azure NetApp Files share and adding it to your datastore because every firm must understand the choices for expanding the datastore in AVS.

Chapter 10, Working with VMware Site Recovery Manager, outlines the process of configuring **Site Recovery Manager (SRM)** between two AVS private clouds. VMware SRM enables you to plan, test, and execute the recovery of VMs between a protected and a recovery vCenter Server site.

Chapter 11, Managing an Azure VMware Solution Environment, demonstrates some best practices for managing your AVS environment. AVS is a VMware-validated solution that is subjected to ongoing verification and testing in order to ensure compatibility with vSphere enhancements and upgrades.

Chapter 12, Leveraging Governance for Azure VMware Solution, clarifies how to leverage governance for your AVS environment using a unified security and compliance approach.

Chapter 13, Summary of Azure VMware Solution, Roadmap, and Best Practices, concludes the book by pointing out some of the key topics that we walked through in the earlier chapters.

To get the most out of this book

To get the most from this book you should already be familiar with VMware solutions and understand Azure networking.

Software/hardware covered in the book	Operating system requirements
ESXi – 7.0 U3c	Windows
VMware vCenter Server – 7.0 U3c	
vSAN – 7.0 U3c	
vSAN on-disk format – 10	
VMware NSX-T Data Center – 3.1.2	
HCX 4.4.2	

Download the color images

We also provide a PDF file that has color images of the screenshots and diagrams used in this book. You can download it here: `https://packt.link/kxOKM`.

Conventions used

There are a number of text conventions used throughout this book.

`Code in text`: Indicates code words in text, database table names, folder names, filenames, file extensions, pathnames, dummy URLs, user input, and Twitter handles. Here is an example: "AVS comes with a built-in user called `cloudadmin` in the new environment's vCenter."

Bold: Indicates a new term, an important word, or words that you see onscreen. For instance, words in menus or dialog boxes appear in **bold**. Here is an example: "Under **Settings**, select **Resource providers**."

> **Tips or important notes**
> Appear like this.

Get in touch

Feedback from our readers is always welcome.

General feedback: If you have questions about any aspect of this book, email us at customercare@ packtpub.com and mention the book title in the subject of your message.

Errata: Although we have taken every care to ensure the accuracy of our content, mistakes do happen. If you have found a mistake in this book, we would be grateful if you would report this to us. Please visit www.packtpub.com/support/errata and fill in the form.

Piracy: If you come across any illegal copies of our works in any form on the internet, we would be grateful if you would provide us with the location address or website name. Please contact us at copyright@packt.com with a link to the material.

If you are interested in becoming an author: If there is a topic that you have expertise in and you are interested in either writing or contributing to a book, please visit authors.packtpub.com.

Share Your Thoughts

Now you've finished *The Ins and Outs of Azure VMware Solution*, we'd love to hear your thoughts! Scan the QR code below to go straight to the Amazon review page for this book and share your feedback or leave a review on the site that you purchased it from.

https://packt.link/r/1801814317

Your review is important to us and the tech community and will help us make sure we're delivering excellent quality content.

Download a free PDF copy of this book

Thanks for purchasing this book!

Do you like to read on the go but are unable to carry your print books everywhere?

Is your eBook purchase not compatible with the device of your choice?

Don't worry, now with every Packt book you get a DRM-free PDF version of that book at no cost.

Read anywhere, any place, on any device. Search, copy, and paste code from your favorite technical books directly into your application.

The perks don't stop there, you can get exclusive access to discounts, newsletters, and great free content in your inbox daily

Follow these simple steps to get the benefits:

1. Scan the QR code or visit the link below

https://packt.link/free-ebook/9781801814317

2. Submit your proof of purchase
3. That's it! We'll send your free PDF and other benefits to your email directly

Part 1: Getting Started with Azure VMware Solution (AVS)

This part provides an introduction to **Azure VMware Solution** (**AVS**) and its architecture. You will also learn about the different use cases where AVS is best suited for an organization.

This part comprises the following chapters:

- *Chapter 1, Introduction to Azure VMware Solution*
- *Chapter 2, Enterprise-Scale for Azure VMware Solution*

1
Introduction to Azure VMware Solution

Azure VMware Solution (**AVS**) is a first-party Microsoft Azure service developed in conjunction with VMware that provides a familiar vSphere-based, single-tenant private cloud on Azure that is like the one used by VMware. The VMware technology stack consists of the following components: vSphere, NSX-T, vSAN, and HCX. AVS is installed on a dedicated infrastructure in Azure data centers and runs natively on that infrastructure. In comparison to existing on-premises VMware infrastructures, AVS provides a consistent and well-known user experience. Customers may deploy an AVS environment in a matter of hours and migrate **Virtual Machine** (**VM**) resources in a matter of minutes. Microsoft supplies all the networking, storage, management, and support services that are required.

The following diagram depicts connectivity between your private cloud (on-premises infrastructure) and Microsoft Azure via an ExpressRoute running your AVS private cloud, as well as other Azure-native services:

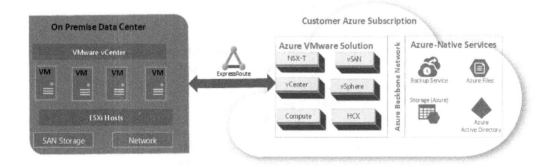

Figure 1.1 – Connectivity relationship between your private clouds and AVS VNets

In this chapter, we're going to cover the following main topics:

- Network connectivity to AVS
- AVS high-level architecture
- Use cases for AVS in the enterprise
- Enterprise-scale for AVS
- Network and connectivity topologies
- Identity and access management
- Business continuity and disaster recovery
- Security, governance, and compliance
- Management and monitoring

Network connectivity to AVS

AVS provides a private cloud environment that can be accessed from both on-premises and Azure-based infrastructure resources. The connectivity is provided by utilizing Azure ExpressRoute, **Virtual Private Network** (**VPN**) connections, or Azure Virtual WAN.

However, to make these services available, specific network address ranges and firewall ports must be configured.

When a private cloud is deployed, private networks are formed for management, provisioning, and vMotion. These private networks will be used to connect to vCenter and NSX-T Manager, as well as to perform virtual machine vMotion and deployment. The private network must use a /22 CIDR notation. This /22 is only used for the management components and not for your workload segments. You will need additional networks for your workloads.

It is possible to link private clouds to on-premises systems using ExpressRoute Global Reach. It establishes direct connections between circuits at the **Microsoft Enterprise Edge** (**MSEE**). Your subscription must have a **Virtual Network** (**VNet**) with an ExpressRoute circuit to on-premises for the connection to work. The reason for this is that VNet gateways (ExpressRoute gateways) are unable to transfer traffic across circuits. This means that you can connect two circuits to the same gateway, but the traffic will not be transferred from one circuit to another.

Each AVS environment is its own ExpressRoute region (and, thus, its own virtual MSEE device), which allows you to connect Global Reach to the "local" peering location by creating a virtual MSEE device for each environment. The ability to connect several AVS instances in a single region to the same peering location is provided by this feature.

AVS hosts, clusters, and private clouds

AVS private clouds and clusters are constructed on top of a hyper-converged Azure infrastructure host. These hosts are dedicated bare metal. At the time of writing, the **High-End (HE)** hosts have 576 GB of RAM and dual Intel 18 Core 2.3 GHz CPUs. In addition, the hosts are equipped with two vSAN disk groups, each of which contains a raw vSAN capacity layer of 15.36 TB (SSD) and a 3.2 TB (NVMe) vSAN cache tier. See the following hardware and software configurations:

AVS Software Specification
ESXi – 7.0U3c Enterprise Plus.
vCenter – 7.0U3c Standard.
vSAN – 7.0U3c Enterprise.
NSX-T – 3.1.2 Datacenter.
HCX Advanced.
HCX Enterprise is also available. Submit a Microsoft support ticket to get an upgrade.

Table 1.1 – AVS software specification

SKU	CPU (GHz)	RAM (GB)	vSAN Cache Tier (TB, Raw)	vSAN Capacity Tier (TB, Raw)	Regional Availability
AV36	Dual Intel 18 Core 2.3 GHz (SkyLake)	576	3.2 (NVMe)	15.20 (SSD)	All product regions
AV36P	Dual Intel 18 Core 2.6 GHz / 3.9 GHz Turbo (Cascade Lake)	768	1.5 (Intel Optane Cache)	19.20 (NVMe)	Selected regions (*)
AV52	Dual Intel 26 core 2.7 GHz / 4.0 GHz Turbo (Cascade Lake)	1,536	1.5 (Intel Optane Cache)	38.40 (NVMe)	Selected regions (*)

Figure 1.2 – AVS hardware SKUs

Creating new private clouds can be done through the Azure site, the Azure CLI, or automated deployment scripts.

There is a minimum of 3 nodes per vSphere cluster, and a maximum of 16 nodes per vSphere cluster, 12 clusters per private cloud instance, and a maximum of 96 nodes per Azure private cloud instance. You can review the Microsoft documentation at this link for more information: `https://docs.microsoft.com/en-us/azure/azure-vmware/concepts-private-clouds-clusters#clusters`.

As you can see from the preceding information, you can scale your private cloud to meet your workload demands.

AVS high-level architecture

AVS provides a private cloud environment that can be accessed from both on-premises and Azure-based infrastructure. Connectivity includes services such as Azure ExpressRoute, VPN connections, and Azure Virtual WAN.

Specific network address ranges and firewall ports, on the other hand, are required for these services to be enabled.

A private cloud is deployed, and private networks are constructed for management, provisioning, and VM movement (*vMotion*).

These private networks will be used to connect to vCenter and NSX-T Manager, as well as for VM vMotion and deployment. You can review the Microsoft documentation at this link for more information: `https://learn.microsoft.com/en-us/azure/azure-vmware/tutorial-network-checklist#routing-and-subnet-considerations`. A connection between private clouds and on-premises settings is made possible through the usage of ExpressRoute Global Reach. Global Reach establishes direct connections between Azure ExpressRoute circuits at the MSEE level. An ExpressRoute circuit to on-premises is required for the connection, which is included in your subscription with a VNet. The reason for this is that VNet gateways (ExpressRoute gateways) are unable to transfer traffic between circuits. This implies that you can connect two circuits to the same gateway, but the traffic will not be transferred from one circuit to the other.

Each AVS environment is deployed with its own 10 GB ExpressRoute circuit (and, thus, its own virtual MSEE device), which allows you to connect Global Reach to the "local" peering location by creating a virtual MSEE device in each environment. It enables you to connect several AVS instances in a single region to the same peering site by using a VNet interface.

See the following high-level AVS networking overview:

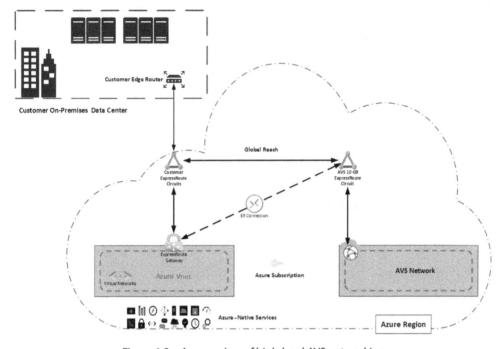

Figure 1.3 – An overview of high-level AVS networking

The preceding diagram shows the logical connections between AVS and the customer's on-premises data center. It also shows the connection between AVS and Azure. Global Reach is used to connect two or more ExpressRoute circuits.

Use cases for AVS in an enterprise

You can migrate your VMware workloads from your on-premises data center to AVS and integrate additional Azure services with ease, using the same VMware tools that you are already familiar with. However, while there are other advantages, we've identified the top five reasons why AVS is proving to be the most cost-effective path to the cloud for many enterprises.

Data center footprint deduction, consolidation, and retirement

Nowadays, we see many customers reducing their on-premises data center footprint for many reasons, including cost, eschewing the management of data centers, and focusing more on their business.

AVS helps customers reduce the size of their data center's footprint by redeploying their VMware-based VMs on a one-time basis.

The vSphere-based workloads can be migrated to AVS in a non-disruptive, automated, scalable, and highly available manner without having to change the underlying vSphere hypervisor.

Data center expansion based on demand

Customers are now able to increase their data center capacity in a seamless and elastic manner, while also adjusting their cost on demand for short periods of time. We see this kind of need in a logistic business, where customers need to increase their data center capacity for a period and then decrease that capacity once it is no longer needed.

Disaster recovery and business continuity

AVS can be used as a primary or secondary on-demand DR site for on-premises data center infrastructure by customers who require a backup data center in the cloud.

Speed and simplification of migration/hybrid cloud

AVS has proven to be one of the most efficient and straightforward methods of getting started on Azure without having to make any changes to your existing apps or servers.

AVS is very cost-effective

When it comes to running VMware apps on Windows Server and SQL Server, AVS is the most cost-effective option. If you use your on-premises data center effectively, you can save money by not having to purchase multiple licenses for both on-premises and cloud applications. When you migrate to AVS, you will receive 3 years of free **Extended Security Updates** (**ESU**) for Windows and SQL Server 2008/2008R2/2012.

Enterprise-scale for AVS

Enterprise-scale for AVS is a collection of open source templates of Azure Resource Manager and Bicep that can be used with AVS planning and deployment. You can think of it as a roadmap for how to build a scalable AVS for future growth. This open source solution gives you an example of how to set up Azure landing zone subscriptions for a scalable AVS. It also gives you an example of how to set up the subscriptions. The architecture and best practices of the Cloud Adoption Framework's Azure landing zones are used in the implementation, with a focus on the design principles of a large-scale deployment.

If you want to make your landing zone more efficient, you should think about how to make it more scalable. It is important for your organization to follow this advice when it comes to making design decisions because this will help it to grow.

There are many ways for people to use AVS, and they all work well. It's possible to use the enterprise-scale option for your AVS set to build a structure that works for you and puts your organization on a path to long-term growth.

To assist you with your AVS setup, enterprise-scale for AVS offers the following resources:

- Customizable environment variables that can be implemented using a modular method
- Helpful recommendations to assess the most important decisions
- A landing zone design that you can use for reference to set up your AVS deployment
- A deployment that includes the following:

    - A reference architecture to deploy your AVS environment
    - A reference architecture approved by Microsoft

Prerequisites for the implementation of the enterprise-scale landing zone for AVS

The AVS construction set is based on the fact that you've already set up an enterprise-scale landing zone. If you want to learn more about enterprise-scale landing zones, check out the following:

- `https://docs.microsoft.com/en-us/azure/cloud-adoption-framework/ready/enterprise-scale/`
- `https://docs.microsoft.com/en-us/azure/cloud-adoption-framework/ready/enterprise-scale/implementation`

There are multiple design guidelines that you will need to go through when creating your landing zone for AVS. The following is a list of areas that you will need to focus on when creating an AVS enterprise-scale landing zone:

- Network and connectivity topology
- Identity and access management
- **Business Continuity and Disaster Recovery (BCDR)**
- Security, governance, and compliance
- Management and monitoring
- Platform automation

Let us dig a bit deeper into these design areas to provide you with some more detailed information.

Network and connectivity topologies

For both cloud-native and hybrid scenarios, implementing a VMware **Software-Defined Data Center** (**SDDC**) with the Azure cloud ecosystem has some unique design challenges to think about when planning for your deployment. Some of these challenges are outlined as follows:

- **Hybrid connectivity**: This is the connectivity between your on-premises environment and your AVS. This is where you will need to look at what connectivity method you are currently using to connect your on-premises data center to Azure if you already have a presence in Azure. If there is no existing connectivity make sure you understand what the options are (ExpressRoute, S2S VPN, or SDWAN). We will dive deeper into these areas in a later chapter.

- **Reliability and performance**: This is very important as you will need to have consistent and low latency for your workloads. You will also need to design for scalability for future growth.

- **A zero-trust network security model**: Security should be the heart of every solution that you implement in Azure, and AVS is no exception. You will need to plan for security for your network perimeter, and for traffic inspection for ingress and egress flows.

- **Extensibility**: Your network footprint should be easily extended without the need for a redesign. This is very important as your AVS needs grow.

We will now review the various network traffic flows within the AVS architecture between AVS, Azure-native services, and a customer's on-premises environment:

- **AVS without any connectivity**:

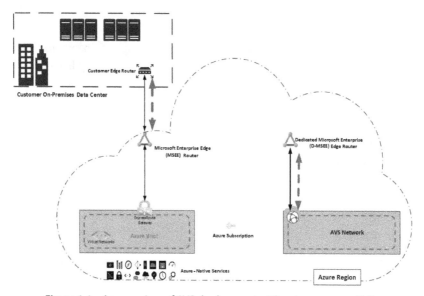

Figure 1.4 – An overview of AVS deployment without any connectivity

The preceding diagram shows AVS deployment without any connectivity to Azure or the customer's on-premises data center.

- **AVS with Global Reach enabled**:

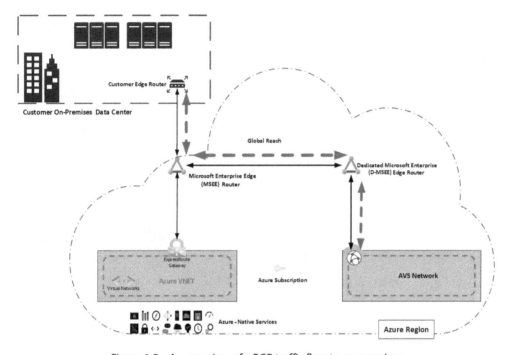

Figure 1.5 – An overview of a BGP traffic flow to on-premises

The preceding diagram shows a BGP traffic flow (blue dotted arrows) from AVS to the customer's on-premises data center. BGP traffic will flow between both environments once Azure Global Reach is enabled.

- **AVS with Global Reach enabled – BGP traffic flowing to Azure from AVS:**

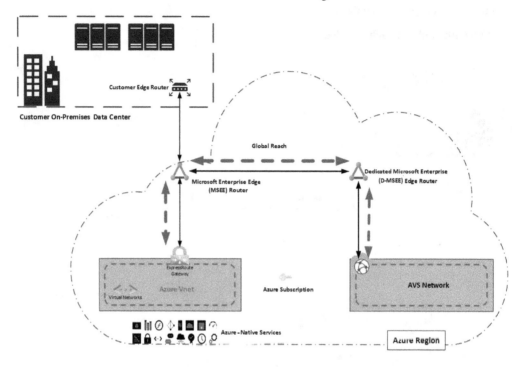

Figure 1.6 – The BGP traffic flow from AVS to Azure-native services through the customer MSEE

The preceding diagram shows the BGP traffic flow from AVS to Azure-native services through the customer's MSEE. BGP traffic will flow between both environments once Azure Global Reach is enabled.

- **AVS connection between AVS and Azure-native:**

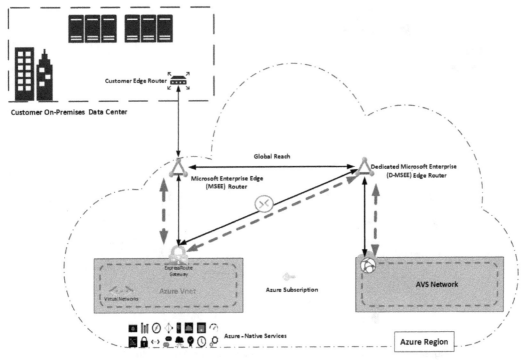

Figure 1.7 – The BGP traffic flow from AVS to Azure-native services
through the customer's ExpressRoute gateway

The preceding diagram shows the BGP traffic flow from AVS to Azure-native services through the customer's ExpressRoute gateway. This connection is only to Azure services and not to the customer's on-premises environment.

- **Internet traffic flow from AVS via a vWAN:**

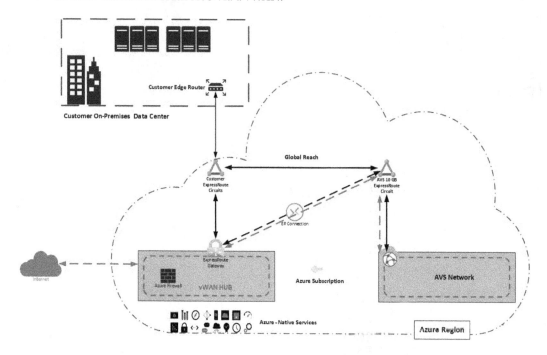

Figure 1.8 – Internet traffic flow from AVS via a secure Azure Virtual WAN

The preceding diagram shows internet traffic flow from AVS via a secure Azure Virtual WAN.

- **Internet traffic flow from AVS via an Azure Route Server and a Network Virtual Appliance (NVA):**

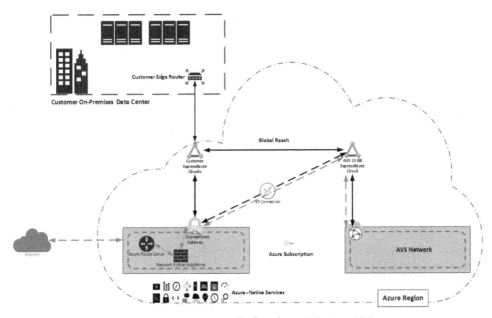

Figure 1.9 – Internet traffic flow from AVS via an NVA

The preceding diagram shows internet traffic flow from AVS via an NVA.

- **Internet traffic flow from AVS via the customer on-premises firewall:**

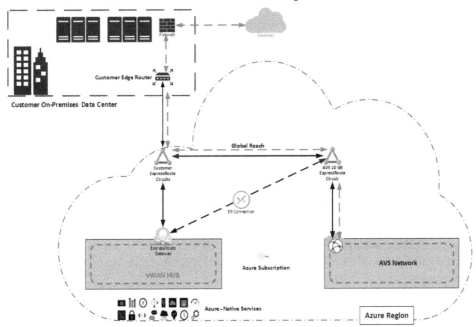

Figure 1.10 – Internet traffic flow from AVS via the customer's on-premises infrastructure

The preceding diagram depicts internet traffic flow between AVS and the customer's on-premises infrastructure, flowing through their firewall.

Identity and access management

There are different identity requirements for AVS based on how it's set up in Azure. AVS comes with a built-in user called `cloudadmin` in the new environment's vCenter. This user has been given the CloudAdmin role, which gives them a lot of power in vCenter. It's also possible to set up new roles in your AVS environment using the principle of least privilege:

- **Active Directory Domain Services (AD DS)**: It is highly recommended to deploy an AD DS domain controller in your identity subscription in Azure. This will help with users' authentication in Azure instead of this request being made back in the customer's on-premises environment.

- **Least-privilege roles**: Allow only a small number of people to have the CloudAdmin role. When assigning users to AVS, use custom roles and as few permissions as possible.

- **Resource-based access control**: People who need to manage AVS should only have **Role-Based Access Control** (**RBAC**) permissions for the resource group where AVS is installed, and for delegated users who need to manage it.

- **vSphere permissions**: Only set up vSphere permissions with custom roles at the top level if you need to. It's better to give permissions to the right **VM** folder or resource pool. In general, do not apply any kind of vSphere permissions at or above the level of the data center.

- **Active Directory sites and services**: Ensure that Active Directory sites and services are configured with the appropriate and respective client IP subnets to provide a better authentication experience when attempting to locate the nearest domain controller.

- **Active Directory groups**: When you set up groups in Active Directory, you can use RBAC to manage vCenter and NSX-T. You can make your own roles and assign them to Active Directory groups.

Business continuity and disaster recovery

Implementing a BCDR solution is very important for all organizations. Businesses need to be able to continue functioning in case of any disruption to day-to-day operations. AVS is no exception.

It is important for an organization and its enterprise application workloads to meet their **Recovery Time Objective** (**RTO**) and **Recovery Point Objective** (**RPO**) goals. Effective BCDR design meets these needs at the platform level. To figure out how to build **DR** capabilities, you need to know what your platform needs.

Even though AVS provides one or more private clouds that have vSphere clusters, built from dedicated hardware, a robust BCDR solution is highly recommended.

Design considerations for AVS BC

Choose a backup solution that has been proven to work for VMware VMs, such as **Microsoft Azure Backup Server (MABS)** or from one of the backup service providers. Some of the backup solutions for AVS are listed as follows:

- **MABS**:

  - When you set up MABS, make sure it is in the same Azure region as your AVS private cloud. This method saves money on traffic costs, makes it easier to manage, and keeps the primary/secondary topology the same.

  - There are two ways to run MABS: you can run it as an Azure VM in your Azure-native environment, or you can run it on an Azure VM within your private cloud. It's very important to put it outside of the AVS private cloud and into a VNet that has connectivity to AVS via ExpressRoute.

  - To get help restoring from a backup for parts of the AVS platform, such as vCenter, NSX Manager, or HCX Manager, you will need to create an Azure support request.

- Cohesity

- Dell Technologies

- Rubrik

- Veritas

- Veeam

- Commvault

Design considerations for AVS DR

The options for designing AVS DR are listed as follows:

- Make sure that the business needs match up with the recovery time, capacity, and recovery point goals for your applications and VM tiers. To make sure you get what you want, plan and design accordingly. Use the right replication technology to do this. Technologies such as SQL always-on availability groups, VMware **Site Recovery Manager (SRM)**, and **Azure Site Recovery (ASR)** are some ideal solutions to implement as part of your DR strategy.

- VMware SRM is a very good option to back up your AVS private cloud to a second AVS private cloud in case of a disaster, so you can keep your business running. Please note that VMware SRM is not included in your AVS subscription. It is an add-on that you will need to have a separate license for.

- **ASR** is another solution that you can use to back up your AVS private cloud to Azure IaaS.

- There are also partner solutions such as JetStream Software that you can use to implement your DR solution for AVS.

- Make sure you decide which of your AVS workloads needs to be protected if there is a DR situation Consider only protecting the things that are important to your business to keep the costs down.

- Make sure to have copies of your domain controllers in your secondary environment.

- Make sure both backend ExpressRoute circuits have ExpressRoute Global Reach turned on. This will make it possible for DR to happen between AVS private clouds in different Azure regions. These circuits connect the main private cloud to the secondary private cloud when DR solutions such as VMware SRM and VMware HCX are used.

Security, governance, and compliance

In this section, we will talk about how to make sure that AVS is safe to use and that you can manage it from start to finish. We will look at some specific design elements and give specific advice for the security, governance, and compliance of your AVS.

Security

It is important to make sure that you have your security components planned out before you deploy any solution in Azure. AVS is no exception. In the following, we will look at some of the key factors to consider:

- **Limits on permanent access**: In the Azure resource group that hosts the AVS private cloud, the Contributor role is used. This role is used by the AVS service. To keep contributor rights from being misused, limit permanent access. Using a privileged account management tool can help you keep track of and limit how long highly privileged accounts are used.

- **Centralized identity management**: AVS gives cloud administrators and network administrators credentials that can be used to set up the VMware environment. They are visible to everyone who has RBAC access to the AVS.

If you want to restrict built-in `cloudadmin` and network administrator users' access to the VMware control plane, use the control plane RBAC features to properly control role and account access. Using least-privilege principles, make a lot of targeted identity objects such as users and groups. Limit access to the administrator accounts provided by AVS and set them up in a *break-glass* configuration. If you can't use any other administrative account, use the built-in account instead.

Use the Cloudadmin account to connect Azure AD DS with the VMware vCenter and NSX-T control applications and the administrative identities for the domain services that are part of the cloud. Use users and groups from your domain to manage and operate your AVS. Don't share your account.

Customize vCenter roles and link them to AD DS groups so that you can control access to VMware control surfaces with fine-grained privilege level control, such as who can see what.

There are options in AVS that you can use to change and reset passwords for vCenter and NSX-T administrators. When you use the break-glass configuration, set up a regular rotation of these accounts, and rotate the accounts when you do.

Governance

Consider following these suggestions when you plan for an environment and guest VM governance:

- **Storage space on your vSAN**: You need to have sufficient free space on your vSAN to maintain your VMware **Service-Level Agreement (SLA)**. A minimum of 25 percent free space on your vSAN is required by VMware.

- **Host quota**: If there are not enough host quotas, there could be delays of up to 7 days before you get more space for growth or DR. Make sure to think about growth and DR when you ask for the host quota, and check the environment's growth and maximums on a regular basis to make sure there is enough time for expansion requests. Suppose a three-node AVS cluster needs three more nodes for DR If you need six nodes, ask for six hosts instead of just the primary three nodes. It doesn't cost extra if you ask for a host quota.

- **Access to the ESXi**: There is limited access to the ESXi hosts. Some third-party software that needs access to the ESXi host might not work. Identify any AVS-supported third-party software in the source environment that needs access to the ESXi host from AVS. Make sure you know how to use the AVS support request process in the Azure portal when you need to get into the ESXi host.

Compliance

There are many recommendations for compliance when planning your AVS environment. A few of these recommendations are listed as follows:

- Monitoring
- Backup
- Country and/or industry regulatory compliance
- Data retention
- Corporate policies

Let us look at compliance in more detail:

- **Microsoft Defender for Cloud monitoring**: When you use Defender for Cloud, you can use the regulatory compliance view to make sure that you are meeting the required security and regulatory standards. Defender for Cloud workflow automation can be set up to keep an eye on how well you're doing in terms of deviation from the required compliance policies.

- **Workload VM backup compliance**: Ensure your AVS guest VMs are being backed up. We mentioned earlier the importance of backing up your AVS in case of a disaster.

- **Country- or industry-specific regulatory compliance**: If you want to avoid costly legal action or fines, make sure your guest workloads for AVS follow local and industry-specific regulations. It's important to know how the cloud-shared responsibility model works for different industrial or regional regulatory compliance.

- **Data retention and residency requirements**: AVS doesn't allow you to keep or get data from clusters that are stored on the cloud. This means that when you delete a cluster, it stops all running workloads and components and also destroys all the cluster's data and settings, such as public IP addresses. You will not be able to recover the deleted data.

- **Corporate policy compliance**: Keep an eye on the guest workloads in AVS to make sure they don't break company rules and regulations. Use solutions such as Azure Arc-enabled servers and Azure Policy, or a similar third-party solution. Routinely check and manage AVS guest VMs and applications to make sure they meet the required internal and external regulations.

Management and monitoring

When planning a public cloud solution, management and monitoring should be integral parts of your design construct, and AVS should be no exception.

Creating an AVS with optimum management and monitoring capabilities will help you get the best out of the solution.

Look at the following tips for managing and monitoring your AVS platform:

- Keep track of the metrics that matter most to your operations teams and make alerts and dashboards that show them.

- vSAN storage space is limited, so you need to keep an eye on vSAN capacity. When you use vSAN storage, only use it for guest VM workloads. VMware requires you to have a minimum of 75 percent free space on the vSAN to maintain the SLA. It is also recommended that you use Azure Blob Storage to store your backups instead of using vSAN storage.

- A local identity provider is used by AVS. After you set up AVS, use a single administrative user account for the first configurations. Active Directory integration is highly recommended, since it provides a way to track the actions of each user.

Summary

AVS is a first-party Microsoft Azure service built in collaboration with VMware that delivers a familiar vSphere-based, single-tenant, private cloud on Azure. The VMware technology stack includes vSphere, NSX-T, vSAN, and HCX. AVS is deployed natively on dedicated infrastructure in Azure data centers. AVS provides a consistent, well-known user experience with existing on-premises VMware environments. Customers can deploy an AVS environment in just a few hours and quickly migrate VM resources. Microsoft provides all necessary networking, storage, management services, and support.

Throughout this chapter, we went over the critical design areas to help you design, implement, secure, and manage AVS.

Some of the critical design areas we covered were as follows:

- AVS overview
- Use cases for AVS
- Enterprise-scale for AVS
- Networking
- Identity and access management
- BC/DR
- Security, governance, and compliance

You should now understand what AVS is and the use cases for the solution.

In the next chapter, we will go deeper into enterprise-scale for AVS and the available guidelines and take a deeper look into the overall architecture.

2
Enterprise-Scale for AVS

As discussed in *Chapter 1*, enterprise-scale for AVS is a collection of open source templates of Azure Resource Manager and Bicep that can be used with AVS planning and deployment. You can think of it as a roadmap for building a scalable AVS for future growth. This open source solution gives you an example of setting up Azure landing zone subscriptions for a scalable AVS environment. It also gives you an example of how to set up the subscriptions. The architecture and best practices of the Cloud Adoption Framework's Azure landing zones are used in the implementation, focusing on the design principles of a large-scale deployment.

In this chapter, we will dive deeper into enterprise-scale for AVS and look at the different design considerations when implementing a scalable AVS in your Azure landing zone.

Customers have multiple options when building out the landing zone. It's possible to use the enterprise-scale option for AVS to build a structure that works for you and puts your company on a path to long-term growth.

A fundamental prerequisite to implementing a successful AVS is that you must have already successfully implemented an Azure enterprise-scale landing zone. It is best to have that in place before deploying AVS, as this will help you connect your AVS infrastructure to your on-premises data center, AVS to Azure, and AVS to the internet.

We will be covering the following topics in depth throughout this chapter:

- Network and connectivity topology for AVS
- Identity and access management for AVS
- Business continuity and disaster recovery

Network and connectivity topology for AVS

As mentioned in the previous chapter, for both cloud-native and hybrid scenarios, implementing a VMware **software-defined data center** (**SDDC**) with the Azure cloud ecosystem has some unique design challenges to think about when planning for your deployment.

Some of the challenges we talked about in the previous chapter are as follows:

- Hybrid connectivity
- Reliability and performance
- A zero-trust network security model
- Extensibility

We will now look at the different networking components and concepts used to create the different connectivity medians for AVS:

- **Azure Virtual Network (VNet)**: This is the building block for your private networks in Azure. When you set up a virtual network, it looks and works like a traditional network running in your own data center. However, it has the benefits of the Azure infrastructure, such as scale, availability, and isolation. When you use Azure VNet, many types of Azure resources, such as virtual machines and databases, can communicate with each other, the internet, and on-premises data centers safely and securely.

- **Hub and spoke network topology**: In this topology, the virtual hub network serves as the central connection point for multiple spoke virtual networks. A spoke virtual network that connects to the hub can be used to separate different types of workloads from each other. An on-premises data center, AVS SDDC, can also be linked up to a hub through a connection point (ExpressRoute and/or a **site-to-site (S2S)** VPN).

- **Network virtual appliance (NVA)**: This is a virtual appliance that provides **Wide-Area Network (WAN)** optimization, security, connectivity to different endpoints, application delivery, and more. Some examples of an NVA include F5-BigIP, Azure Firewall, Cisco Firewall, Barracuda Firewall, and others. An NVA in Azure functions the same way a physical appliance does in a customer data center.

- **Azure Virtual WAN (vWAN)**: vWAN is a unique networking service that you can use to integrate many features such as networking, routing, and security to provide a single interface for operation.

Some of the functionalities of Azure vWAN include S2S VPN connectivity, ExpressRoute connectivity, which is a private connection, routing, and Azure Firewall. It also includes encryption for private connectivity. You can start with just one use case, and then add functionalities as they are needed.

The architecture for Azure vWAN is a hub and spoke architecture that can scale as needed by adding additional spokes:

- **Layer 4 (L4)**: The fourth layer of the OSI model is referred to as layer 4. It is also known as the **transport layer**. L4 enables data to be transmitted or transferred between hosts or end systems transparently. Error recovery and flow control are both handled by L4. The following are some of the protocols used in L4:

- **Transmission Control Protocol (TCP)**

- **Multipath TCP (MPTCP)**

- **User Datagram Protocol (UDP)**

- UDP-Lite

- **Reliable UDP (RUDP)**

- **AppleTalk Transaction Protocol (ATP)**

- **Sequenced Packet Exchange (SPX)**

- **Layer 7 (L7)**: The application layer, L7, is the final layer of the OSI model and is the highest layer. Layer 7 identifies the communication parties and the level of service between them. It is L7's job to keep data private and authenticate users, and it does so by looking for any limits on the data syntax. This layer is responsible for all API interactions. The following are some of the main protocols of L7:

  - HTTP

  - HTTPS

  - SMTP

Understanding networking requirements for AVS

Setting up the landing zone for AVS requires a thorough understanding of Azure network design and implementation techniques. A wide range of capabilities is supported by Azure networking products and services. How to arrange services and choose the right architecture relies on your organization's workloads, governance, and requirements since every organization is different.

Here, you will find some essential requirements and considerations that will affect your AVS deployment decision:

- Connectivity from on-premises data centers to AVS, where you will be connecting over ExpressRoute or an S2S VPN. Will ExpressRoute Global Reach be enabled?

- Will AVS be connecting to an Azure VNet hub for connectivity to Azure native services or a vWAN hub?

- Is there an L2 extension from the on-premises data center to AVS (this is done to retain a VM's IP addresses)?

- Do you have an NVA in your current Azure environment?

- Will applications require HTTP/S or not for internet ingress?

- Traffic inspection needs the following:

  - AVS access to Azure native services

  - AVS access back to the on-premises data center

  - Egress access to the internet from AVS

  - Ingress from the internet to AVS

  - Traffic connection with AVS

Networking scenarios for AVS with traffic inspection

Let us look at four scenarios!

Scenario 1 – a secure vWAN hub with default route propagation enabled

This scenario will be ideal for customers for whom the following applies:

- Do not need traffic inspection between AVS and their on-premises data center

- Do not need any traffic inspection between AVS and their Azure vNet

- Do need the traffic to be inspected between AVS and the internet

In this scenario, the customer will have to add services for L4 and L7 ingress if they so require. We are also assuming that the customer already has an ExpressRoute connection in place between their on-premises data center and Azure.

You can implement this architecture with the following components:

- An application gateway for L7 load balancing and SSL offloading, which will reside in a spoke VNet

- An Azure Firewall in the secure vWAN hub (or any other NVA)

- L4 **destination network address translation** (**DNAT**) must be configured on the Azure Firewall to filter and translate ingress network traffic

- Configure all egress traffic through the Azure Firewall in the vWAN hub.

- Implement ExpressRoute, SD-WAN, or a VPN connection between AVS and the on-premises data center

The following diagram illustrates scenario 1:

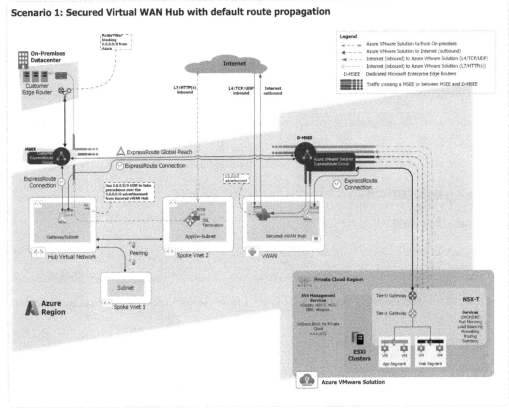

Figure 2.1 – Secure vWAN hub with default route propagation enabled

Things to consider

If the default 0.0.0.0/0 route that is being advertised from AVS is interfering with your existing environment, you will need to take additional steps to prevent route propagation.

The default 0.0.0.0/0 route from the vWAN hub propagates to the ExpressRoute gateway. It takes precedence over the internet system route built into the virtual network if you currently connect to a hub-and-spoke topology-based virtual network via an ExpressRoute gateway rather than directly. If this is an issue, a workaround is to create a 0.0.0.0/0 user-defined route on the virtual network to override the default route learned.

Azure Firewall advertises the 0.0.0.0/0 route to the VMware solution in the secured vWAN hub. Advertisement of the 0.0.0.0/0 route is routed to the customer's on-premises environment through Global Reach. Implement an on-premises route filter to stop the 0.0.0.0/0 route from being learned. If an SD-WAN or VPN is implemented, this behavior won't happen.

Any VPN, ExpressRoute, and virtual network connections to the vWAN hub that don't need the 0.0.0.0/0 advertisement will also get the advertisement, even though they don't need it. To solve this problem, you can use an on-premises edge device to block the 0.0.0.0/0 route.

You could also do the following:

- Disconnect the ExpressRoute, VPN, or virtual network from the secured vWAN hub

- Enable 0.0.0.0/0 propagation

- Disable 0.0.0.0/0 propagation on those specific connections where you do not need the 0.0.0.0/0 route

- Reconnect those connections once the 0.0.0.0/0 route has been disabled

Scenario 2 – egress from AVS with either NSX-T or NVA

This scenario will be ideal for customers if:

- The customer will use the built-in NSX-T solution

- Customers need to have traffic inspection done in AVS and will bring their own NVA solution

- ExpressRoute is currently or planned to be in place from the customer's on-premises environment into Azure

- L4 or HTTP/S ingress from the internet will be needed

There is a single point of entry for all traffic between AVS, Azure Virtual Network, the internet, and on-premises data centers: the NSX tier-0/tier-1 router or the provided NVAs.

You can implement this architecture with the following components:

- An NVA or NSX **Distributed Firewall (DFW)** behind the tier-1 gateway in AVS

- An application gateway for L7 load balancing and SSL termination, which will reside in a spoke VNet

- Azure Firewall for L4 DNAT. Azure Firewall will reside in a secured vWAN hub.

- An ExpressRoute solution deployed between the on-premises environment and Azure with Global Reach enabled.

The following diagram illustrates scenario 2:

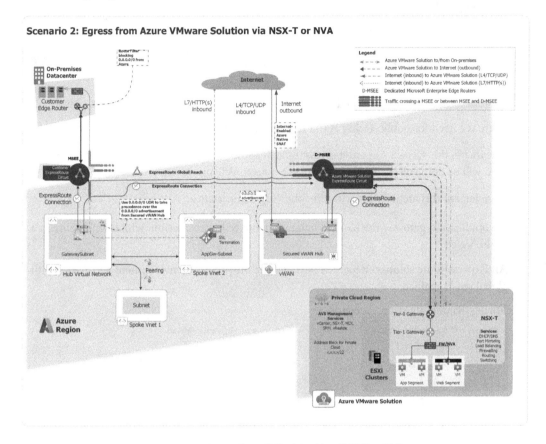

Figure 2.2 – Egress from AVS with either NSX-T or NVA

Things to consider

Internet access will need to be enabled from the Azure portal. Another key thing to keep in mind is that the IP address for internet access is dynamic. The public IP address is not coming from the NVA but a managed service in AVS.

Since the NVA is not a part of the AVS solution by default, the customer will need to bring their own license, and it is the customer's responsibility to implement high availability for the NVAs.

Scenario 3 – AVS's egress and ingress network traffic is routed through an on-premises firewall

This scenario will be ideal for customers if the following applies:

- The customer wants to use a firewall that resides in the on-premises environment the 0.0.0.0/0 route is being advertised from

- ExpressRoute is already deployed or planned to be deployed with Global Reach enabled

- There is a need for public-facing HTTP/S access or L4 ingress services

Please note that in this scenario, internet egress traffic inspection is handled on-premises. The secured vWAN hub will be doing all the traffic inspection between AVS and Azure vNet.

You can implement this architecture with the following components:

- An application gateway for L7 load balancing and SSL termination, which will reside in a spoke VNet

- An ExpressRoute solution deployed between the on-premises environment and Azure with global reach enabled

The following diagram illustrates scenario 3:

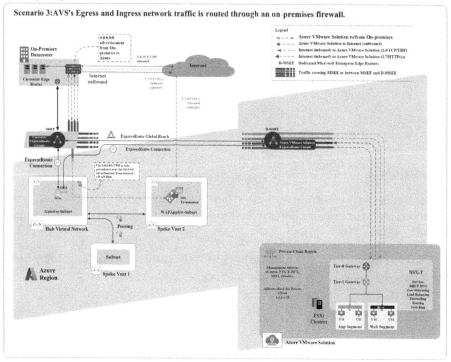

Figure 2.3 – AVS's egress and ingress network traffic is routed through an on-premises firewall

Things to consider

In this scenario, the public IP address for internet access resides on the NVA, which is on-premises.

The default 0.0.0.0/0 route from the vWAN hub propagates to the ExpressRoute gateway. It takes precedence over the internet system route built into the virtual network if you are currently connected to a hub-and-spoke topology-based virtual network via an ExpressRoute gateway rather than directly. If this is an issue, a workaround is to create a 0.0.0.0/0 user-defined route on the virtual network to override the default route learned.

Scenario 4 – all network traffic is inspected by third-party NVAs in the hub vNet

This scenario will be ideal for customers if the following applies:

- Global reach is not enabled due to restrictions or policy
- Your ExpressRoute circuit does not support global reach
- Customers need to have more control over their firewalls deployed in Azure
- Customers need to continue using the NVAs that they currently use to have a seamless firewall experience
- The need for traffic inspection between AVS and your on-premises environment using the NVAs in Azure

This scenario takes into consideration that ExpressRoute is already deployed between the customer's on-premises environment and Azure.

You can implement this architecture with the following components:

- Hosted third-party NVAs in VNets for firewalls and other networking functionalities, including L7 load balancing and SSL termination
- Azure Route Server (Route Reflector), which will be used to route traffic between AVS, Azure VNets, and the on-premises data center

This scenario also takes into consideration that global reach is not enabled, so the NVAs will now be responsible for internet access and routing back to the on-premises environment from AVS.

Also, please note that this is a very complex configuration, so use another option if Global Reach is available to you.

The following diagram illustrates scenario 4:

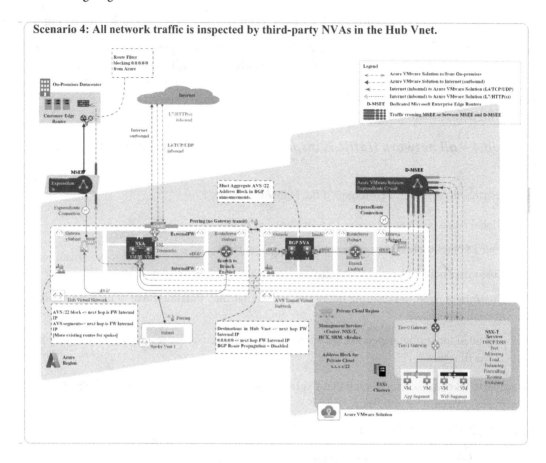

Figure 2.4 – All network traffic is inspected by third-party NVAs in the hub vNet

Things to consider

Since global reach is disabled, you will need to make sure that you inspect all traffic in the hub VNet where the NVAs are hosted.

To make sure that all traffic is routed through the hub VNet, deploy Azure Route Server. Since the NVAs are a third-party solution, you are responsible for the implementation and management of that solution.

It is also recommended to implement the NVAs in high availability for redundancy purposes.

Recommendations for AVS networking design

The following are the recommendations for the AVS networking design:

- Internet, ExpressRoute, HCX, public IP, and ExpressRoute Global Reach are shared by all clusters. Some basic networking parameters, such as network segments, the **Dynamic Host Configuration Protocol** (**DHCP**), and the **Domain Name System** (**DNS**), can also be shared throughout application workloads.

- Before deployment, plan out your private clouds and clusters. Your networking requirements are directly affected by the number of private clouds. For private cloud management and IP address segment for VM workloads, each private cloud requires its own /22 address space. Consider setting aside some time to define those address spaces.

- It is very important to discuss with your networking team how to segment and distribute your private clouds, clusters, and network segments for workloads. This will help you avoid wasting IP addresses. Do this during the planning phase of your architecture.

- Because all clusters share the same /22 address space, they can communicate within an AVS private cloud.

- Use NSX's built-in DHCP service or a private cloud's local DHCP server for DHCP. Don't send broadcast DHCP traffic back to on-premises networks across the WAN.

- Deploy a new DNS infrastructure in your AVS private cloud if you are not connecting back to an on-premises environment (Windows Server DNS services running on an Azure VM or Azure Private DNS: `https://learn.microsoft.com/en-us/azure/dns/private-dns-overview`).

- If the AVS infrastructure is connected to an on-premises environment, use your existing DNS solution. If needed, deploy DNS forwarders to extend AVS.

Identity and access management for AVS

We recognize that AVS identification needs differ, depending on the AVS implementation in Azure; as a result, we'll focus on some of the most typical instances:

- There are different identity requirements for AVS based on how the solution will be utilized. AVS comes with a built-in user called cloudadmin in the new environment's vCenter. This person has been given the CloudAdmin role, which gives them a lot of power in vCenter. It's also possible to set up new roles in your AVS environment using the principle of least privilege.

- Limit RBAC permissions for AVS in Azure to the Resource Group where it's installed and the number of users who need to maintain it.

- To manage vCenter and NSX-T, create groups in **Active Directory** (**AD**) and use RBAC. You can create custom roles and assign them to the AD groups.

- The administrator has access to the vCenter `administrator@vsphere.local` account in an on-premises vCenter and ESXi implementation. They can also be allocated to more AD users and groups.

- The administrator does not have access to the administrator user account in an AVS deployment. They can, however, use vCenter to assign AD users and groups to the CloudAdmin role. The CloudAdmin role cannot add an identity source to vCenter, such as an on-premises LDAP or LDAPS server. However, you can add an identity source and assign the CloudAdmin role to users and groups using `run` commands.

> **Note**
>
> The private cloud user has no access to or control over the management components that Microsoft supports and controls. This includes clusters, hosts, data stores, and distributed virtual switches.

vCenter privileges

On your AVS private cloud vCenter, you can see the privileges that have been assigned to the Azure AVS CloudAdmin role by following these steps:

1. Sign into your vSphere Client and go to **Menu | Administration**.

2. Under **Access Control**, click on **Roles**.

3. Select **CloudAdmin** from the list of roles, then select **PRIVILEGES**:

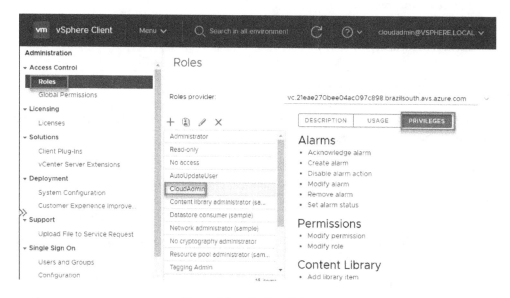

Figure 2.5 – vCenter privileges

> **Information**
>
> For more information on the CloudAdmin privileges for vCenter, please see `https://docs.vmware.com`.

Creating a custom role in vCenter

Custom roles are supported in AVS. These roles can have equal or fewer privileges than the default CloudAdmin role.

The CloudAdmin role is used to create, edit, and delete custom roles with privileges that are less than or equal to those of their current role. You can establish roles with higher rights than CloudAdmin, but you can't assign them to users or groups, and you can't delete them.

It is recommended to clone the CloudAdmin role as the foundation for new custom roles to avoid producing roles that can't be assigned or deleted.

How to create a custom role in vCenter

The steps are as follows:

1. You will need to sign in to vCenter Server with `cloudadmin@vsphere.local` or a user with the CloudAdmin role.
2. Click on **Menu | Administration | Access Control | Roles**.
3. Select **CloudAdmin** and then select the **Clone role action** icon:

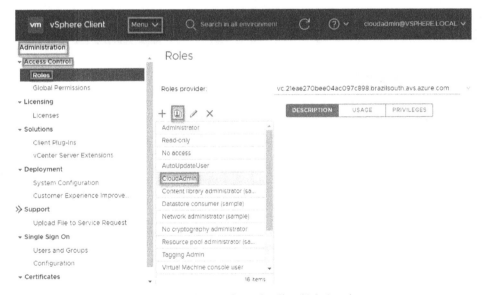

Figure 2.6 – How to clone the CloudAdmin role

4. Enter the name you want for the new clone role:

Figure 2.7 – Creating a new CloudAdmin user

5. You can add or remove roles and then select **OK**. The newly cloned role will now be visible in the **Roles** list:

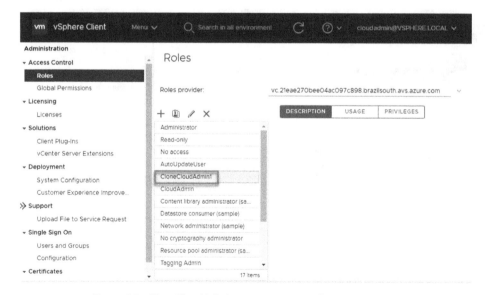

Figure 2.8 – New CloudAdmin user account under the Roles tab

You have now created your custom role in vCenter!

Business continuity and disaster recovery

AVS makes it possible to have one or more private clouds that have vSphere clusters in them. These clusters are built on bare-metal Azure servers. There must be at least 3 ESXi hosts in each cluster, and there can be up to 16 hosts in each cluster. As many as 96 hosts can be run in a private cloud at the same time. The solution comes with vCenter Server, vSAN, vSphere, and NSX-T as part of the private cloud offering. A 10-gigabit ExpressRoute is also a part of the `default solution`.

As robust as the solution is by default, a business continuity and disaster recovery solution is highly recommended.

Design considerations for business continuity in AVS

The following are the design considerations for business continuity in AVS:

- Make sure to use a backup solution that has been certified for AVS. The following is a list of backup solutions that have been certified for AVS:

  - **Microsoft Azure Backup Server (MABS)**

  - Cohesity

  - Dell Technologies

  - Rubrik

  - Veritas

  - Veeam

  - Commvault

- In AVS, the storage policies for VMware vSAN are set up to ensure that the storage is available. When there are three to five hosts in a cluster, the number of host failures that can happen without losing data is one. Two hosts can go down before data is lost if the cluster has between 6 and 16 hosts. VMware vSAN storage policies can be set up for each virtual machine. This means that each virtual machine can have its own storage policy. When you use VMware VMs, you can change the policy used to meet your own needs.

- On AVS, VMware **high availability (HA)** is enabled by default. The HA admittance policy guarantees that a single node's compute and memory capacity is reserved. This reservation ensures that there is enough reserve capacity in an AVS cluster to resume workloads in another node.

The recommended design considerations for business continuity in AVS

The following are the recommended design considerations for business continuity in AVS:

- Use a backup solution to back up your AVS environment.

- Install the backup solution in the same Azure region as the AVS solution. This form of deployment lowers traffic costs, simplifies management, and maintains the primary/secondary topology.

- MABS is available as an Azure **Infrastructure-as-a-Service (IaaS)** VM or within the AVS infrastructure. It's highly recommended to deploy it outside of the Azure VMware solution's private cloud and in a separate Azure VNet. This is because vSAN is a limited capacity resource within AVS. This virtual network is connected to the same ExpressRoute to reduce vSAN consumption.

The following diagram illustrates the recommended design considerations for business continuity in AVS:

Microsoft Azure Backup Server recommended design for AVS

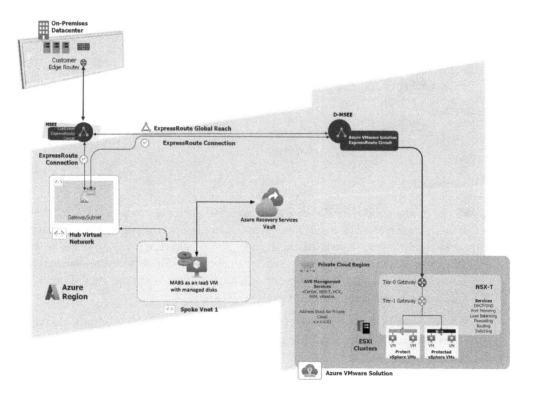

Figure 2.9 – MABS deployment recommendations for AVS

Design considerations for AVS disaster recovery

In the previous chapter, we look at some design considerations to think about when planning your BC/DR solution for AVS. We will now look at a few more things to keep in mind:

- If you are using Azure Site Recovery, make sure that the correct startup order for each workload is listed in your recovery plan. An example will be to have your Domain Controllers start first, followed by your database server, then your application servers.

- When working with disaster recovery, you can use the same IP address spaces from the primary Azure region in the secondary Azure region. However, it requires extra overhead to create the solution:

  - **Keep the same IP addresses**: On the recovered VM, you can utilize the same IP addresses that were assigned to the AVS VMs. Create segregated VLANs or segments in the secondary site for this strategy, and make sure none of these isolated VLANs or segments are connected to any other environment. Change your disaster recovery routes to reflect the new IP address locations and the subnet's relocation to the secondary site. While this strategy works, when looking for little involvement, it causes engineering overhead.

  - **Use new IP addresses**: For restored VMs, you can also utilize alternate IP addresses. The custom IP map will be detailed in the VMware Site Recovery Manager recovery plan if the VM is moved to a secondary location. To update your IP address, select this map. The new IP address is assigned to a configured virtual network when using Azure Site Recovery.

- You must enable ExpressRoute Global Reach between both backend ExpressRoute circuits to enable disaster recovery between AVS private clouds in different Azure regions. When disaster recovery solutions such as VMware SRM and VMware HCX are required, these circuits provide primary to secondary private cloud communication.

Know the difference between a partial and a full disaster recovery solution.

VMware SRM can be used for both partial and full disaster recovery. You can fail some or all the VMs from primary to secondary regions when operating AVS in regions 1 and 2.

If partial or full disaster recovery is possible, it depends on the need for VM recovery and IP address retention.

Design considerations for AVS disaster recovery

When working with AVS on both primary and secondary sites, use VMware Site Recovery Manager. Protected and recovery sites are terms used to describe primary and secondary sites, respectively. The following diagram depicts a high-level overview of vSphere replication in continuous mode:

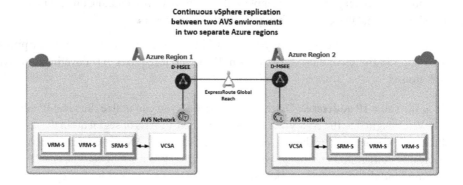

Figure 2.10 – SRM continuous replication, high-level architecture

The secondary and primary site components are depicted in greater depth in the following diagram:

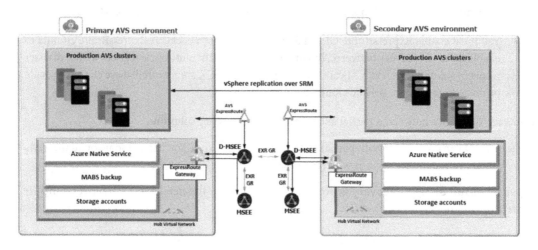

Figure 2.11 – SRM continuous replication, detailed architecture

Connect the primary and secondary AVS private clouds using ExpressRoute Global Reach.

Reduce manual input by incorporating automated recovery plans into each of the solutions. These plans are useful when disaster recovery for the AVS private cloud is provided by VMware Site Recovery Manager or Azure Site Recovery. For failover, a recovery plan organizes machines into recovery groups. This aids in defining a methodical recovery procedure by allowing small autonomous units to be created that can fail over.

If Azure IaaS is the disaster recovery target for AVS, use Azure Site Recovery.

Summary

In this chapter, we looked at the main components of enterprise-scale for AVS. Enterprise-scale for AVS is an open source set of Azure Resource Manager and Bicep templates for planning and deploying AVS. You may consider it a template for building a scalable AVS that can scale up in the future. This open source solution explains how to construct a scalable AVS environment using Azure landing zone subscriptions. It also uses an example to demonstrate how to set up subscriptions. With a focus on large-scale deployment design concepts, the implementation follows the architecture and best practices of the Cloud Adoption Framework's Azure landing zones.

By now, you should understand the critical designs for each of the topics covered. Here is a recap of what was discussed throughout:

- Setting up the landing zone for AVS requires a thorough understanding of Azure network design and implementation techniques.
- It's also possible to set up new roles in your AVS environment using the principle of least privilege.
- AVS makes it possible to have one or more private clouds that have vSphere clusters in them.
- On AVS, VMware HA is enabled by default.
- The HA admittance policy guarantees that a single node's compute and memory capacity is reserved.
- Use a backup solution to back up your AVS environment. Install the backup solution in the same Azure region as the AVS solution.
- MABS is available as an Azure IaaS VM or within the AVS infrastructure.
- It's highly recommended to deploy it outside of the Azure VMware solution's private cloud and in a separate Azure VNet.
- Connect the primary and secondary AVS private clouds using ExpressRoute Global Reach.
- Reduce manual input by incorporating automated recovery plans into each of the solutions.
- These plans are useful when disaster recovery for the AVS private cloud is provided by VMware Site Recovery Manager or Azure Site Recovery.
- If Azure IaaS is the disaster recovery target for AVS, use Azure Site Recovery.

In the upcoming chapter, we will look at planning for a successful AVS deployment.

Part 2: Planning and Deploying AVS

For a successful production-ready environment for building **virtual machines** (**VMs**) and migration, planning your AVS deployment is crucial. You will learn about the necessary steps to plan a successful AVS environment.

This part comprises the following chapters:

- *Chapter 3, Planning for an Azure VMware Solution Deployment*
- *Chapter 4, Deploying Your First Azure VMware Solution Cluster*
- *Chapter 5, Deploying and Configuring HCX in Azure VMware Solution*
- *Chapter 6, Adding Network Segments in Azure VMware Solution*

3

Planning for an Azure VMware Solution Deployment

Planning your **Azure VMware Solution** (**AVS**) deployment is critical for a successful, production-ready environment for building **virtual machines** (**VMs**) and migration. Throughout the planning process, you'll identify and collect the various pieces of information you'll need for your deployment. Please note the data you collect as you plan so that you can refer to it during the deployment. After a successful deployment, you'll have a production-ready environment for creating and migrating VMs.

Throughout this chapter, we will go over the following topics to ensure you get all the required information for a successful AVS deployment:

- Subscription identification
- Resource group identification
- Azure region
- AVS resource name
- VMware HCX
- Why use VMware HCX?
- HCX appliance IP requirements

Subscription identification

One of the things you will need to do is to identify the subscription that you will be deploying AVS in. You can use an existing subscription or create a new one for AVS.

An Azure subscription is a logical grouping of Azure services associated with an Azure account. You will need to have a subscription to use Azure's cloud-based services since it acts as a single billing unit for the Azure resources used by that account:

- An Azure subscription is linked to a single account used to create the subscription and is used for billing purposes. A subscription can contain numerous resources.

- You can have many subscriptions for various reasons, including billing, because each subscription creates its own set of billing reports and invoices.

- Separate subscriptions can also be used to create a division of duty for Azure services. The person who creates an Azure subscription becomes the global administrator for that subscription and has full access to every part of it. This is a standard technique that many organizations follow.

The following illustration is a hierarchy that outlines the management and separation between subscriptions and the respective Azure resources:

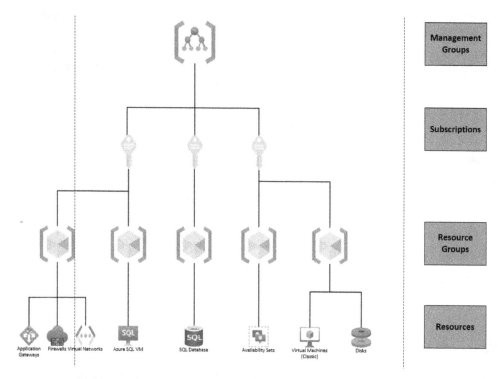

Figure 3.1 – Azure subscription flow

Resource group identification

After the subscription is identified, you will need to decide on the resource group. You can use an existing resource group or create a new one specifically for AVS.

A resource group is a container for Azure solutions that houses connected resources. The resource group can contain all the solutions resources or just the ones you want to manage as a group. Based on what makes the most sense for your company, you select how to allocate resources to resource groupings. Add resources with the same lifetime to the same resource group to make it easier to publish, update, and delete them.

The metadata for the resources is stored in the resource group. As a result, when you specify a location for the resource group, you're also specifying the location of the metadata. You may need to verify that your data is stored in a specific location for compliance reasons.

Azure region

An Azure region consists of multiple data centers to provide redundancy and availability of your applications. You create Azure resources in defined geographic regions such as *East US*, *North Central US*, or *West US*. This approach gives you flexibility as you design applications to create VMs closest to your users and to meet any legal, compliance, or tax purposes.

It is possible to have multiple resources communicating with each other in different regions. However, it is highly recommended that all resources for AVS be deployed in the same region.

Region pairs in Azure

Within the same geography, each Azure region is associated with another. This strategy provides for resource replication across geography, such as VM storage, which should lessen the likelihood of natural disasters, civil unrest, power failures, or physical network outages hitting both regions simultaneously. Region pairs also have the following advantages:

- In the event of a more significant Azure outage, one region from each pair is prioritized to help speed up application recovery
- To minimize downtime and the possibility of an application outage, planned Azure updates are rolled out one by one to paired regions

You can see the full list of Azure regional pairs at `https://docs.microsoft.com/en-us/azure/virtual-machines/regions`.

AVS resource name

The resource name, for example, `ABCPrivateCloud1`, is a descriptive name for your AVS private cloud.

It's critical to note that the name can't be more than 40 characters. You won't be able to create public IP addresses for usage with the private cloud if the name exceeds this limit.

Host size

The AVS clusters are built on bare-metal, hyper-converged infrastructure. At the time of writing this book, the host choices are AV36, AV36P, and AV52. The hosts' RAM, CPU, and disk capacity are listed as follows:

Host Type	CPU (GHz)	RAM (GB)	vSAN Cache Tier (TB, raw)	vSAN Capacity Tier (TB, raw)	Network Interface Cards	Regional availability
AV36	Dual Intel Xeon Gold 6140 CPUs with 18 cores/CPU @ 2.3 GHz, Total 36 physical cores (72 logical cores with hyperthreading)	576	3.2 (NVMe)	15.20 (SSD)	4x 25 Gb/s NICs (2 for management & control plane, 2 for customer traffic)	All product regions
AV36P	Dual Intel Xeon Gold 6240 CPUs with 18 cores/CPU @ 2.6 GHz / 3.9 GHz Turbo, Total 36 physical cores (72 logical cores with hyperthreading)	768	1.5 (Intel Cache)	19.20 (NVMe)	4x 25 Gb/s NICs (2 for management & control plane, 2 for customer traffic)	Selected regions (*)
AV52	Dual Intel Xeon Platinum 8270 CPUs with 26 cores/CPU @ 2.7 GHz / 4.0 GHz Turbo, Total 52 physical cores (104 logical cores with hyperthreading)	1,536	1.5 (Intel Cache)	38.40 (NVMe)	4x 25 Gb/s NICs (2 for management & control plane, 2 for customer traffic)	Selected regions (*)

Figure 3.2 – AVS node specification

Determining the number of hosts and clusters

The first AVS deployment you'll do is a private cloud with just one cluster. You'll need to specify the number of hosts you wish to deploy to the first cluster for your deployment.

Clusters can be added, removed, and scaled. By default, one vSAN cluster is established for each private cloud. Three nodes is the minimum for an AVS cluster.

It is highly recommended that an assessment be done in your environment to determine the VM count, CPU usage of each VM, and storage usage. There are different tools that you can use for this assessment. Azure Migrate, Movere, and RVTools are examples of assessment tools that you can use.

Once the assessment is done, work with your Microsoft account team, where they will do a node count exercise and figure out pricing.

Most other cluster configuration and operation aspects are handled by vSphere and NSX-T Manager. vSAN oversees all local storage on each host in a cluster.

Host quota request for AVS

AVS is not enabled in your Azure subscription by default. Because of this, you need to make a request for a host quota from the Azure subscription that you will use to deploy AVS. Give yourself up to 5 business days for the AVS quota to be enabled for the specified Azure subscription.

The steps for how to request an AVS host quota are as follows:

1. Log in to the Azure portal, then select the subscription in which you want to deploy AVS.
2. In the subscription, scroll down to the bottom of the page, under **Help + Support**, and then create a new support request.
3. On the **Problem description** tab, enter the following information:

   - **Issue type: Technical**
   - **Subscription**: Select your subscription
   - **Service: All services | Azure VMware Solution**
   - **Resource: General question**
   - **Summary: Need capacity**
   - **Problem type: Capacity Management Issues**
   - **Problem subtype: Customer Request for Additional Host Quota/Capacity**

See the following screenshot for additional guidance:

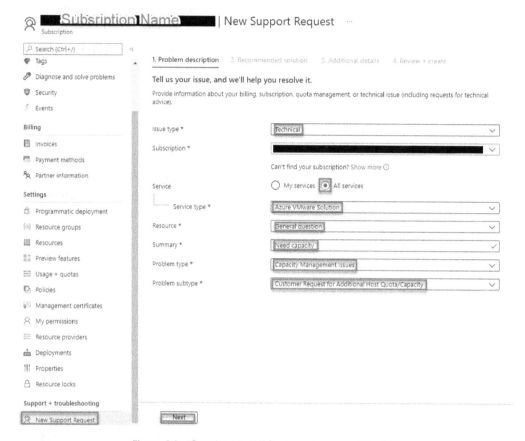

Figure 3.3 – Creating an AVS host quota request (part 1)

4. Click **Next**. On the **Recommended solution** tab, click **Next**.

5. On the **Additional details** tab, under **Problem details**, do the following:

    - Enter the current date and time.

    - In the **Description** box, type in the region in which you will be deploying AVS and enter the number of nodes needed:

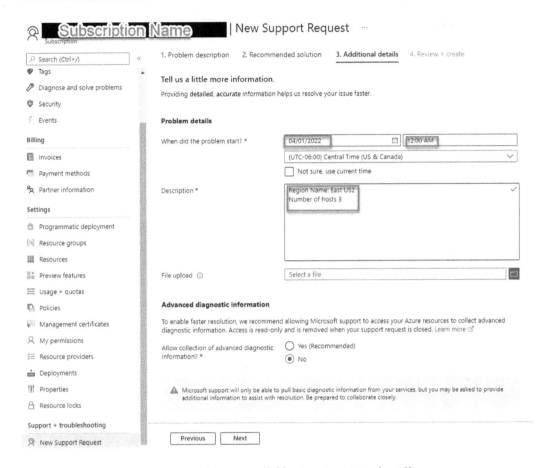

Figure 3.4 – Creating an AVS host quota request (part 2)

6. Scroll down to the **Support method** section and select your preferred contact method:

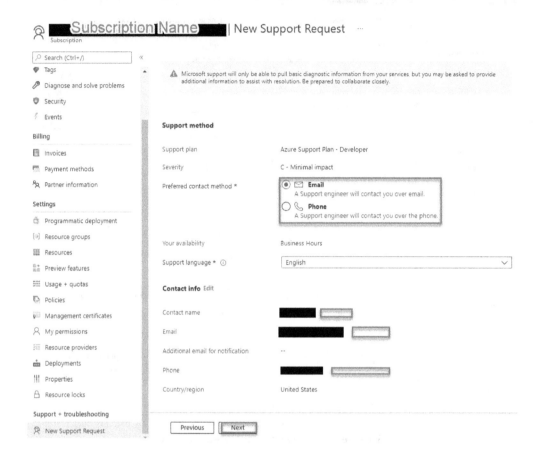

Figure 3.5 – Creating an AVS host quota request (part 3)

7. Select **Email** or **Phone**. Your contact info will auto-populate from your Azure credentials.

8. Click **Next**.

9. On the **Review + create** page, click **Create**.

Requesting a /22 CIDR IP segment for AVS management components

A /22 CIDR network, such as 50.0.0.0/22, is required for AVS deployment. This address space is divided into smaller network segments (subnets) for AVS administration, such as vCenter Server, VMware HCX, NSX-T Data Center, and vMotion. The following diagram shows the IP address segments for AVS management:

Microsoft Azure Backup Server recommended design for AVS

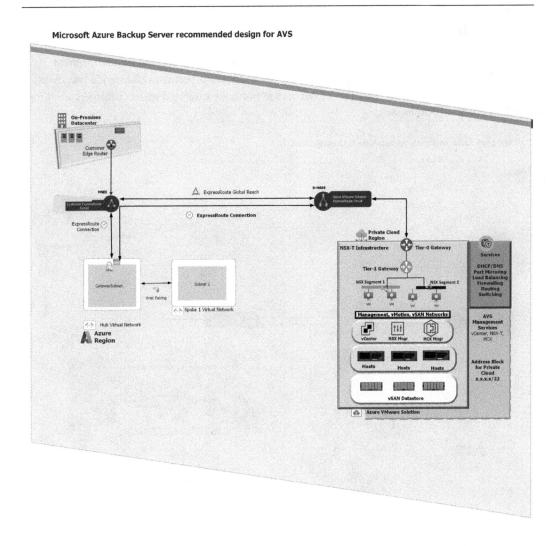

Figure 3.6 – AVS management segments

Please note that any current network segment, on-premises or Azure, should not overlap with the /22 CIDR network.

> **Information**
>
> For a list of the AVS management subnets, go to https://docs.microsoft.com/ EN-us/azure/azure-vmware/tutorial-network-checklist#routing- and-subnet-considerations.

Defining the AVS workload network segments

The VMs must connect to a network segment, just like any other VMware vSphere environment. As AVS's production deployment grows, it's common to see a mix of on-premises L2 extended segments and local NSX-T network segments. The L2 network is normally extended when customers want to retain their current IP addresses.

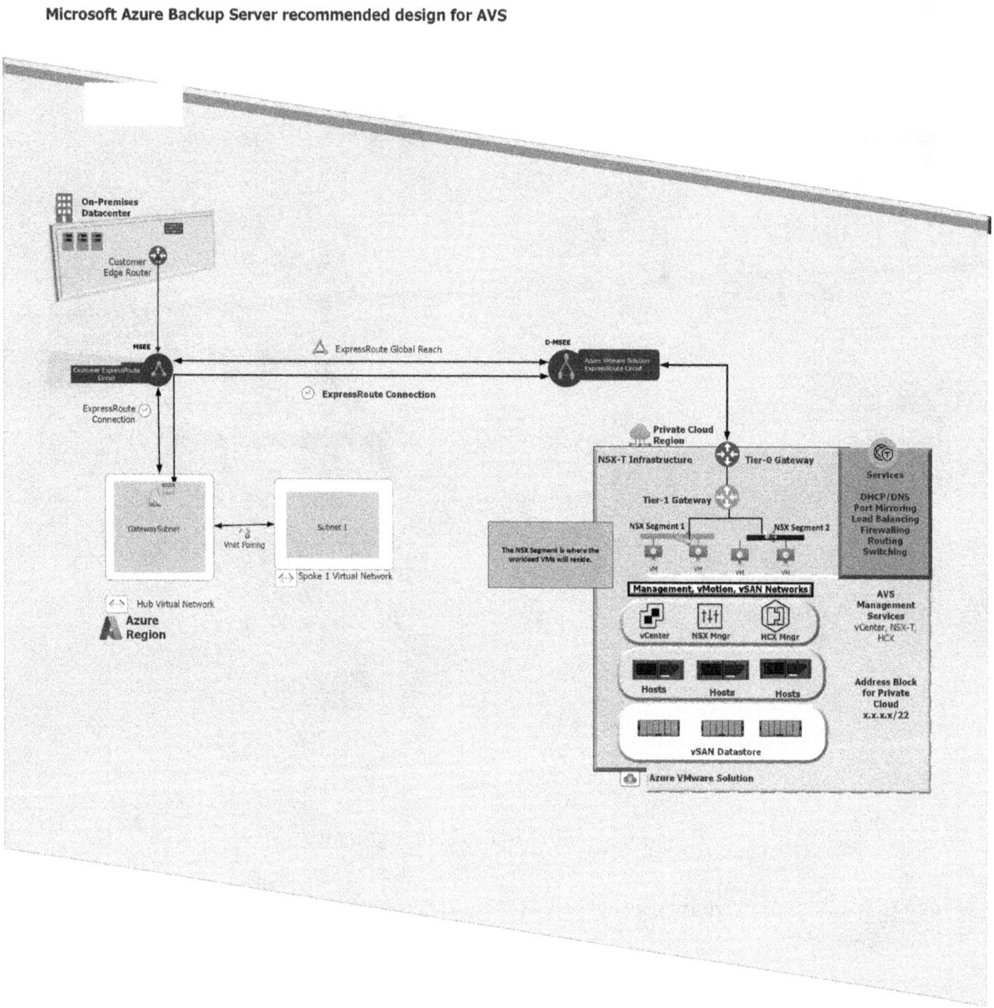

Microsoft Azure Backup Server recommended design for AVS

Figure 3.7 – AVS workload segments

Determine a single network segment (IP network) for the initial deployment, such as 10.0.2.0/24. During the first deployment, this network section is mainly utilized for testing. The address block must not overlap with any network segments on-premises or in Azure, and it must not be within the already specified /22 network segment.

Defining the virtual network gateway

AVS can be connected to an Azure site-to-site VPN connection. However, because of its dedicated connectivity and minimal latency, an ExpressRoute circuit is strongly recommended. The following diagram illustrates an ExpressRoute Gateway connection:

Microsoft Azure Backup Server recommended design for AVS

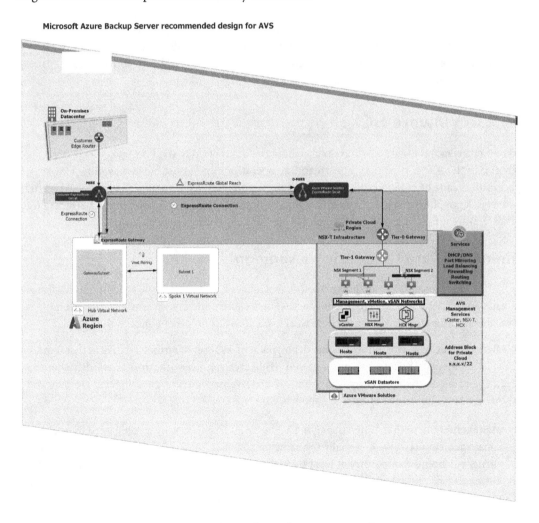

Figure 3.8 – ExpressRoute Gateway connection

To use the ExpressRoute circuit, you'll also need an Azure Virtual Network Gateway. Determine whether you want to use an existing ExpressRoute virtual network gateway or a new ExpressRoute virtual network gateway. It's OK to use an existing ExpressRoute virtual network gateway, but keep track of which ExpressRoute virtual network gateway you'll use for planning purposes. You can choose between three ExpressRoute Gateway SKUs: Standard, HighPerformance, and UltraHighPerformance are the three options. These gateways have throughput speeds of 1 gigabit, 2 gigabits, and 10 gigabits, respectively.

VMware HCX

VMware HCX isolates VMware vSphere-based on-premises and cloud resources and provides them to apps as a single continuous resource. The development of a network extension is automated using an encrypted, high-throughput, WAN-optimized, load-balanced, traffic-engineered, hybrid connection. This enables hybrid services such as application migration, workload balancing, and disaster recovery optimization. Applications may run everywhere with a VMware HCX hybrid connection in place, regardless of the hardware and software beneath.

Why use VMware HCX?

As your company moves toward a hybrid cloud architecture based on AVS, you'll set up new environments locally and in the AVS environment to streamline operations and boost business agility. However, a hybrid cloud can't be fulfilled unless these new environments have applications and workloads. VMware HCX simplifies the process of filling and continuously improving application placement on current VMware infrastructure, enabling data center and cloud conversions.

Defining VMware HCX network segments

VMware HCX is an application mobility technology that makes it easier to migrate applications, rebalance workloads, and maintain business continuity across data centers and clouds. Various migration types are available for migrating VMware vSphere workloads to AVS and other connected sites.

The VMware HCX Connector automates the deployment of a subset of virtual appliances that require multiple IP segments. The IP segments are used while creating network profiles. Modify them as needed depending on the needs of your migration. Determine which of the following are required for a VMware HCX implementation that supports your use case:

- **Management network**: When deploying VMware HCX on-premises, you'll need to establish a management network. It's usually the same management network as your VMware vSphere cluster on-premises. Identify at least two IPs for VMware HCX on this network segment. Depending on the extent of your rollout beyond the pilot or small use case, you may require more IP addresses.

- **Uplink network**: You'll need to identify an Uplink network for VMware HCX when deploying it on-premises. Use the same network that you'll be using for management.

- **vMotion network**: When deploying VMware HCX on-premises, you must designate a vMotion network for VMware HCX, which is typically the same network your on-premises VMware vSphere cluster uses for vMotion. Identify at least two IPs on this network segment for VMware HCX; more numbers may be required depending on the extent of your deployment beyond the pilot or small use case.

  The vMotion network must be exposed on a distributed virtual switch or vSwitch0. If it isn't, change the surroundings to make it so.

- **Replication network**: A replication network must be defined when deploying VMware HCX on-premises. Use the same network as for your Management and Uplink networks. If the on-premises cluster hosts use a dedicated Replication VMkernel network, reserve two IP addresses in this network segment and use the Replication VMkernel network for replication.

HCX appliance IP requirements

The following are the HCX appliance IP requirements:

- Five IP addresses for the HCX appliance

- Four management IPs:

  - One for the HCX manager

  - One for the Interconnect appliance

  - Two will be used for the Network Extension appliance

- One vMotion IP for the Interconnect appliance to participate in vMotion

HCX port requirements

You will need to make sure that all required ports (`https://ports.esp.vmware.com/home/VMware-HCX`) are open for communication between the on-premises components and AVS.

Please see the following diagram for the HCX port requirements:

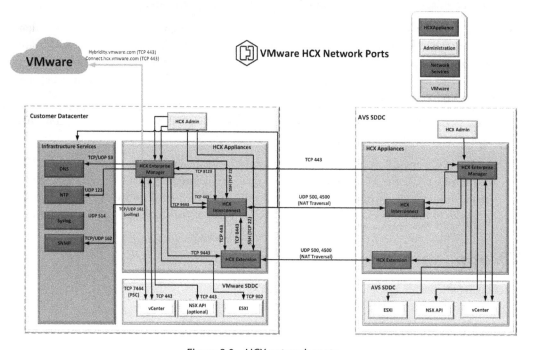

Figure 3.9 – HCX network ports

Benefits of extending the L2 network to AVS

HCX **Network Extension** (**NE**) enables the extension of a broadcast domain from a customer's on-premises data center to AVS through a **Layer 2 VPN** (**L2VPN**). HCX NE functionality is delivered via a dedicated virtual appliance at both locations.

Customers' ability to expand their L2 network is a great option. Customers elect to extend their L2 network during VM migration to AVS since re-IPing these VMs may not be possible. Once the customer's network segment or data center has been thoroughly hydrated, the gateway should be moved to AVS. This should be used as a stopgap measure until the migration is complete.

Prerequisites for extending the L2 network to AVS

Some prerequisites need to be in place before you can extend your L2 network to AVS:

- VMware NSX-T 3.0 or above on AVS

- For expanding vDS-based networks, on-premises **vSphere Distributed Switch** (**vDS**) version 5.1.0 or above is necessary

- VMware HCX deployment and service mesh are up and running

- Extending NSX-T-based networks requires VMware NSX-T 2.4 or above

L2 extension recommendations or considerations

The following are the L2 extension recommendations:

- Never extend networks used for VMware HCX network profiles, vSphere management networks, or other VMkernel networks (for example, vMotion/vSAN). The reason is that network loops or IP/MAC conflicts are not detected or mitigated by VMware HCX NE.

- It is not supported to extend vSphere Standard Switch-based networks. However, you can extend VMware NSX networks. To expand NSX networks, the NSX manager must be registered with HCX.

- A maximum of three AVS private clouds may be added to a network.

- Up to 8 networks may be extended using a single piece of VMware HCX NE equipment. Up to 128 HCX NE appliances may be managed using HCX manager 4.x.

- In an extended L2 network, the default gateway is left in the customer's data center. This can sometimes lead to routing issues due to the latency between the default gateway on-premises and the migrated VM in AVS. This situation may be addressed using HCX **Mobility Optimized Networking** (**MON**).

Planning your L2 network extension

VMware HCX NE connects a client location to an AVS private cloud through a Layer 2 VPN. This service is completely integrated into HCX and performs comparable functions to the NSX L2 VPN. Using an alternate bridging option, such as NSX L2 VPN, with HCX NE is not supported. Customers should choose a single L2 extension solution that will suit their migration or disaster recovery requirements.

HCX NE appliances are installed in pairs, with one at the source site and the other at the destination location. UDP ports 500 and 4500 are used for the encrypted tunnel between HCX NE appliances. If there are any firewalls in the route between the appliances, they should be set to enable communication on these ports.

Customers should be aware of the benefits and drawbacks of adopting HCX NE as an optional service. There are alternatives to utilizing HCX NE, such as assigning new IPs to VMs as they move or migrating a whole network with all associated VMs to the cloud in one go. When none of these methods are practicable, HCX NE is a useful tool. While the HCX NE appliance is built for dependability and rapid startup, it is not intended for high availability (vSphere High Availability can be used to mitigate this concern). In addition, beginning with HCX 4.0, HCX 4.X features an in-service upgrade option for HCX NE appliances, reducing the downtime from a software update to a matter of seconds.

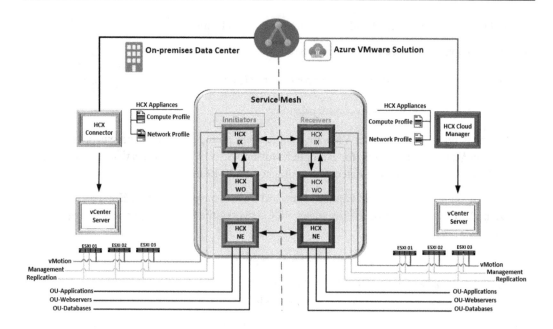

Figure 3.10 – HCX Network Extension layout

Using HCX NE in conjunction with other HCX services may improve the performance and traffic flow. TCP Flow Conditioning is performed by HCX Traffic Engineering, which dynamically changes MSS to decrease fragmentation in NE traffic. HCX Mobility Optimized Networking optimizes traffic flows for VMs that have been migrated to AVS and are connected to an extended network.

Summary

Throughout this chapter, the focus was on the importance of planning for your new AVS deployment. For a successful production-ready environment for creating new VMs and migrating existing workloads, planning your AVS deployment is crucial. You'll identify and gather the many pieces of information you'll need for your deployment during the planning phase. After a successful deployment, you'll have a production-ready environment for creating and migrating VMs.

Some of the critical topics that needed to be identified were as follows:

- Azure subscription

- Resource group

- Azure region

- Resource name

- Host size

- Determining the number of hosts and clusters
- Requesting a host quota for an eligible Azure plan
- Requesting a /22 CIDR IP segment for private cloud management from your networking team
- Defining the AVS workload network segments
- Defining the virtual network gateway

You would be able to deploy an AVS environment with the preceding information if this was a greenfield without any migration from on-premises using VMware HCX. However, if you plan to use VMware HCX for vMotion, you must plan your VMware HCX deployment and configuration.

In the next chapter, we will be looking at deploying your first AVS cluster.

4
Deploying an Azure VMware Solution Cluster

Now that we have been introduced to AVS and planning for an AVS deployment, it's time to do an actual deployment.

AVS gives you the ability to deploy a vSphere cluster in Azure within hours instead of weeks, as is typical when deploying a vSphere cluster on-premises. For each private cloud created, there's one vSAN cluster by default. You can scale up and scale down as needed. The minimum number of hosts per cluster is three. More hosts can be added 1 at a time, up to 16 hosts per cluster. The maximum number of clusters per private cloud is 12. The initial deployment of AVS has three hosts.

After completing this chapter, you will have learned about the requirements needed to deploy an AVS cluster. You will also have learned how to validate a deployment and connect to your Azure environment and on-premises data center using different Azure networking solutions.

Before we go through the actual steps to deploy an AVS cluster, we want to ensure that you have what you need for the deployment to be seamless and successful.

Throughout this chapter, we will cover the following topics:

- Prerequisites to deploy AVS
- AVS deployment validation
- Connecting AVS to your Azure infrastructure
- Connecting AVS to your on-premises environment

Prerequisites to deploy AVS

In this section, we will cover the various prerequisites that you will need to deploy AVS.

Registering the AVS resource provider

One of the first things you will need to do is register the AVS resource provider (Microsoft.AVS) in the subscription to which you will be deploying AVS. This is required to enable all AVS-related features and functions in the specified subscription.

To register the AVS resource provider, please follow these steps:

1. Log in to the Azure portal.

2. Select the subscription that AVS will be deployed to.

3. Under **Settings**, select **Resource providers**.

4. In the search box on the right-hand side of the page, type `Microsoft.AVS`.

5. Make sure **Microsoft.AVS** is selected and click **Register**:

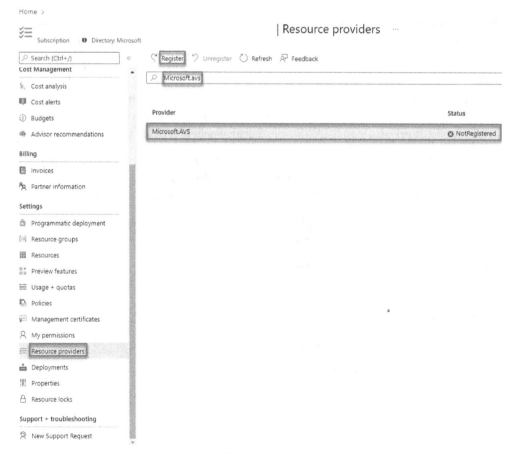

Figure 4.1 – Microsoft.AVS resource provider registration

Registration takes a few minutes to be completed.

6. The **Microsoft.AVS** resource provider should now have a **Registered** status:

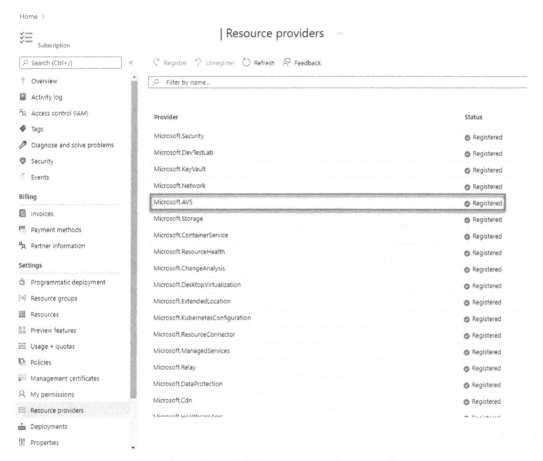

Figure 4.2 – Microsoft.AVS resource provider registration

Deploying an AVS cluster

Before deploying a successful AVS cluster, you will need to ensure that a few housekeeping items are in place. See the following screenshot for more information:

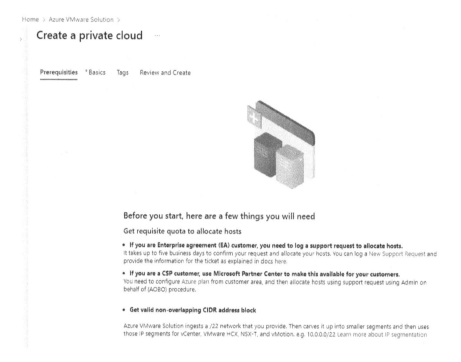

Figure 4.3 – The AVS deployment checklist

The basic information needed to deploy AVS

Now that the resource provider is registered and you have the AVS host quota approved for your subscription, it's time to make sure you have the following information ready:

Item	Value	Example
Subscription	Select the name of the subscription where you will be deploying AVS.	Corp-Infra-AVS-Sub
Resource group	Select the name of the resource group you identified in the AVS planning phase.	Corp-Infra-AVS-RG

Resource name	Provide a name for your AVS private cloud. You should have already decided on a name in the planning phase.	Corp-Infra-SDDC
Location	For location, select the region you decided on for AVS deployment.	East US
Size of hosts	Currently, there's only one host size for AVS. It's AV36.	AV36
Host location	The options for this are around availability zones. Currently, AVS is only deployed to a single availability zone.	All host in one availability zone.
Number of hosts	Based on the number of nodes you decided on during the assessment phase, this is where you will enter that. The minimum number of nodes for an AVS cluster is three. You can add additional nodes as needed.	3
Address block	This is the /22 CIDR block that was identified in the planning phase. This is used for the management components of AVS. You will still need additional networks for your workload segments.	50.0.0.0/22

Table 4.1 – AVS basic information

For more details, refer to the following screenshot:

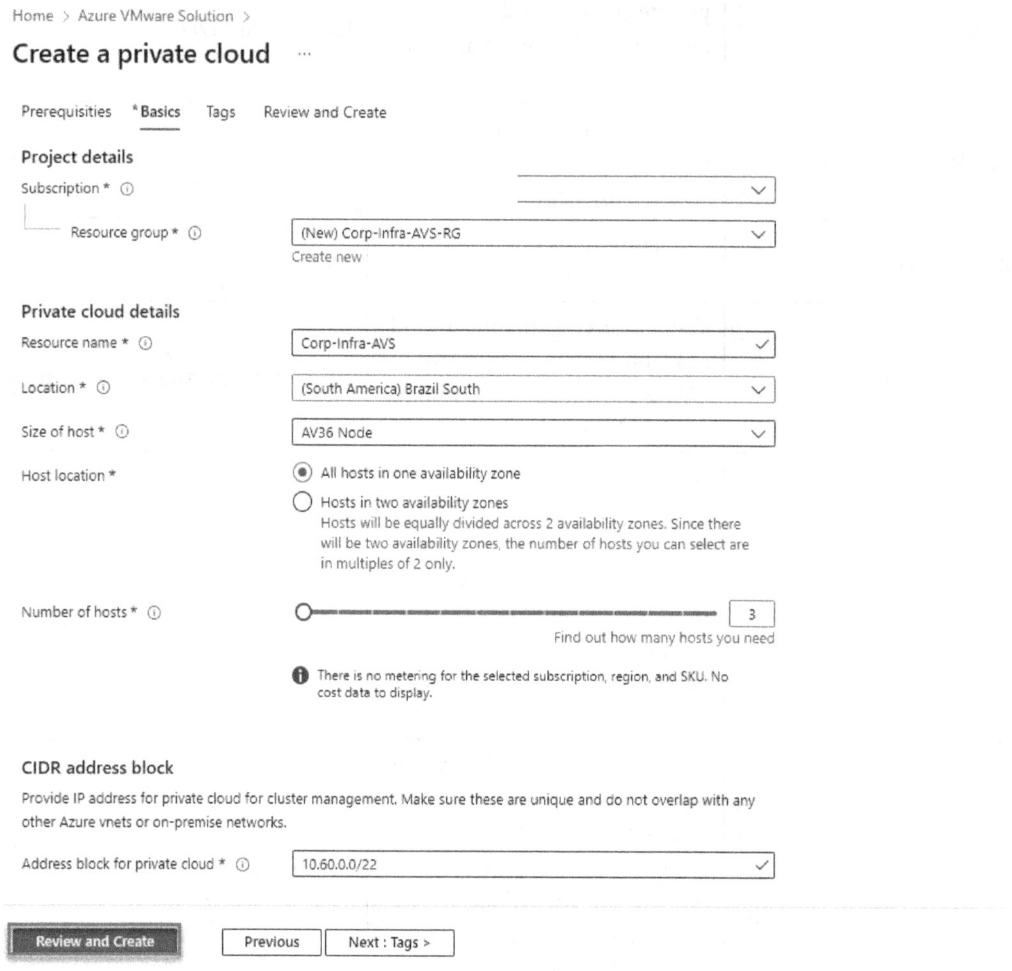

Home > Azure VMware Solution >

Create a private cloud ...

Prerequisities   *Basics   Tags   Review and Create

Project details

Subscription * ⓘ

Resource group * ⓘ   (New) Corp-Infra-AVS-RG
Create new

Private cloud details

Resource name * ⓘ   Corp-Infra-AVS

Location * ⓘ   (South America) Brazil South

Size of host * ⓘ   AV36 Node

Host location *   ⦿ All hosts in one availability zone

   ◯ Hosts in two availability zones
     Hosts will be equally divided across 2 availability zones. Since there
     will be two availability zones, the number of hosts you can select are
     in multiples of 2 only.

Number of hosts * ⓘ   O————————————————————— [3]
     Find out how many hosts you need

ⓘ There is no metering for the selected subscription, region, and SKU. No
cost data to display.

CIDR address block

Provide IP address for private cloud for cluster management. Make sure these are unique and do not overlap with any other Azure vnets or on-premise networks.

Address block for private cloud * ⓘ   10.60.0.0/22

[Review and Create]   [Previous]   [Next : Tags >]

Figure 4.4 – The AVS Basics page

After you have filled in all the required information, click on the **Review and Create** button. This will now take you to the validation page:

Figure 4.5 – AVS deployment legal terms

Once validation is passed, you will have an opportunity to review the legal terms. After reviewing the legal terms, click on **Create** to start the deployment of AVS. This process takes 3 to 4 hours to be completed.

AVS deployment validation

To validate that the AVS deployment was successful, navigate to the resource group that you deployed AVS into. Select **AVS Private cloud** and click on **Overview**. The status of **Succeeded** will be displayed after a successful deployment:

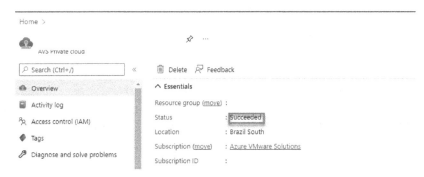

Figure 4.6 – AVS deployment validation

Now that you have a successful deployment of AVS, it is now time to connect it to your Azure environment.

Connecting AVS to your Azure infrastructure

Let's walk you through the steps to connect AVS to an existing ExpressRoute virtual network gateway:

1. The first steps is requesting an ExpressRoute authorization key. Navigate to the new AVS private cloud in the Azure portal. Select **Manage** | **Connectivity** | **ExpressRoute** and then click on + **Request an authorization key**:

Figure 4.7 – Requesting an ExpressRoute authorization key

2. Enter a name for the key and then select **Create**. It may take about 30 seconds to create the key. After the key is created, it will appear in the list of authorization keys for your private cloud:

Figure 4.8 – The ExpressRoute authorization key

3. Now, copy the authorization key and the ExpressRoute ID. These will be needed to complete the peering.

4. Navigate to the ExpressRoute virtual network gateway that you have decided to use and then select **Connections | + Add**:

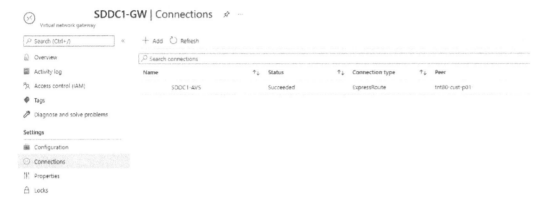

Figure 4.9 – Creating an ExpressRoute gateway connection

5. On the **Add connection** page, provide the required values for the different fields, and then select **OK**:

Field	Value
Name	Enter a name for this connection
Connection type	Select **ExpressRoute**
Redeem authorization	Make sure to check this box
Virtual network gateway	Select the virtual network gateway you decided to use
Authorization key	Paste the authorization key you copied before
Peer circuit URI	Paste the ExpressRoute ID you copied before

Table 4.2 – ExpressRoute virtual network gateway connection field and values

The screenshot for your reference is as follows:

Figure 4.10 – ExpressRoute virtual network gateway connection fields

You will now have a successful connection between your ExpressRoute circuit and your virtual network:

Figure 4.11 – ExpressRoute gateway successful connection

You should now be able to connect between the Azure virtual network where the ExpressRoute circuit is terminated and AVS.

Validating the connection between AVS and Azure

We will now validate the connection between AVS and Azure. The steps are as follows:

1. Use a **jumpbox** (VM) in the Azure virtual network where AVS ExpressRoute is connected.

2. Log in to the Azure portal.

3. Select the VM that you planned on using for the connection. Go to **Settings | Networking** and then select the network interface:

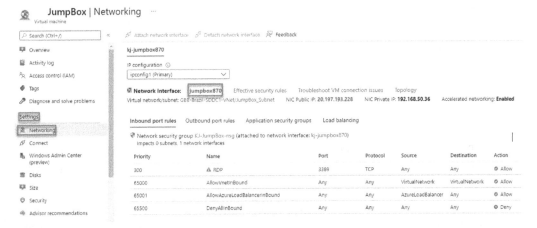

Figure 4.12 – Jumpbox network connection

4. Next to **Network Interface**, select **Effective security rules**; you'll see a list of address prefixes that are contained within the /22 CIDR block you entered during the deployment phase.

5. To connect to vCenter and NSX-T Manager in AVS, open a web browser and enter the IP addresses for both vCenter and NSX-T. You will need to do this from the jumpbox VM that is running in Azure within the same VNet AVS ExpressRoute is connected to.

6.   You will be able to get both IP addresses from the AVS portal page. Go to **Manage | Identity**. Copy the **Web client URL**, **Admin username**, and **Admin password** values for both vCenter and NSX-T Manager:

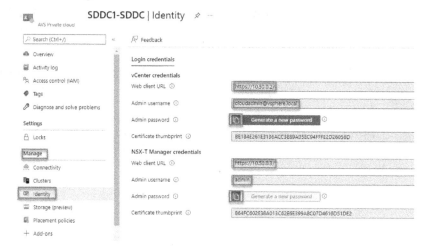

Figure 4.13 – AVS vCenter and NSX-T Manager credentials

The IP address for vCenter will always be the first IP address from the /22 network that you used for the AVS management during your deployment phase. For example, if the /22 network was 10.50.0.0/22, the vCenter IP address will be 10.50.0.2.

7.   Once you have a browser window opened on the jumpbox in Azure, enter the IP address you copied for vCenter. When VMware vSphere is presented, paste the username and password you copied earlier for vCenter:

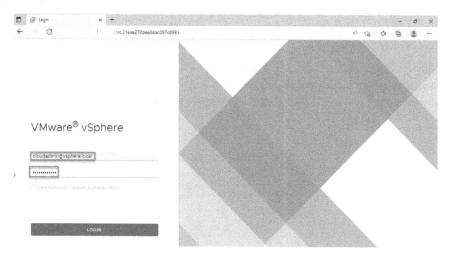

Figure 4.14 – The AVS vCenter login page

8.   Click **LOGIN**. You will now be presented with the vSphere Client in AVS:

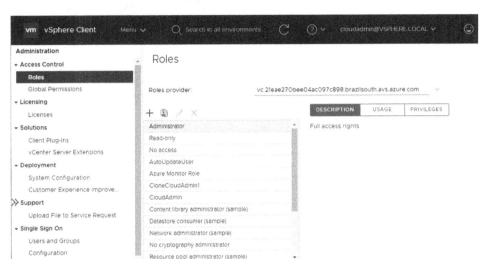

Figure 4.15 – The AVS vSphere Client

The IP address for NSX-T Manager will always be the second IP address from the /22 network that you used for the AVS management during your deployment phase. For example, if the /22 network was 10.50.0.0/22, the NSX-T IP address will be 10.50.0.3.

9.   Repeat *step 6* to log in to NSX-T Manager.

Figure 4.16 – The AVS NSX-T login page

10. Click **LOG IN**. You will now be presented with the vSphere Client in AVS:

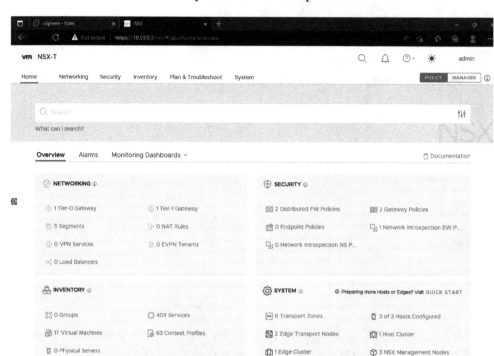

Figure 4.17 – AVS NSX-T Manager

Now that you have connected AVS to Azure, let us see how to connect AVS to your on-premises environment in the next section.

Connecting AVS to your on-premises environment

Connecting the AVS environment to your on-premises data center is needed for virtual machine migrations.

ExpressRoute Global Reach is used to peer multiple ExpressRoute circuits together. In this case, you will use it to establish the connection between the AVS ExpressRoute circuit (DMSEE) and your existing ExpressRoute circuit (MSEE).

Creating an ExpressRoute authorization key on your on-premises ExpressRoute circuit

To create an ExpressRoute authorization key, follow these instructions:

1. In the Azure portal, navigate to the ExpressRoute page. Under **Settings**, select **Authorizations**.

2. Enter a name for the new authorization key and select **Save**.

3. Copy the newly created authorization key and the ExpressRoute ID (resource ID):

Figure 4.18 – Requesting an authorization key from an on-premises circuit

Note that it may take about 30 seconds to create the key. Once the key is created, it will appear in the list of authorization keys for the circuit.

Peering AVS with your on-premises environment

With the new authorization key and ExpressRoute ID from your on-premises circuit, you can peer both circuits together.

Go to the AVS portal, and under **Manage**, select **Connectivity | ExpressRoute Global Reach | Add**:

Figure 4.19 – AVS to on-premises peering with Global Reach

Enter the ExpressRoute ID and the authorization key you created earlier, and click **Create**. The connection will be displayed in the **On-prem cloud connections** list:

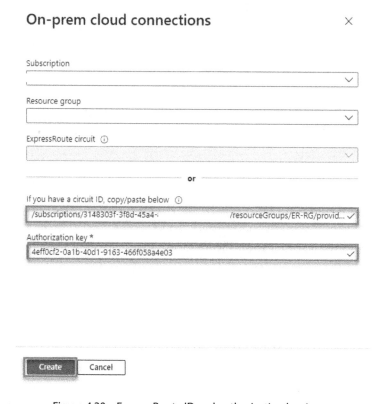

Figure 4.20 – ExpressRoute ID and authorization key input

Validating the connection between AVS and on-premises

In your on-premises edge router, you should now see the management segment from your AVS NSX-T network.

You should also be able to connect from a server or workstation in your on-premises environment to AVS.

Open a browser on your server or workstation from on-premises and go to the vCenter IP address that's in your AVS portal:

Figure 4.21 – AVS vCenter and NSX-T Manager credentials

See *step 6* under the *Validating the connection between AVS and Azure* section for detailed instructions on connecting to AVS vCenter and NSX-T Manager.

Summary

This chapter focused on the prerequisites and steps needed to deploy and connect your AVS environment to Azure and your on-premises data center. Planning for your AVS deployment is critical, as it allows you to ensure that all the information you need will be ready before the actual deployment.

The key areas that were covered in this chapter were the following:

- Prerequisites to deploy AVS
- AVS deployment validation
- Connection validation to both Azure and on-premises

In the next chapter, we will be deploying and configuring HCX, which is needed for connectivity to your on-premises VMware environment. HCX is also needed for virtual machine migration between the on-premises VMware environment and AVS.

5

Deploying and Configuring HCX in Azure VMware Solution

Azure VMware Solution (**AVS**) no longer comes with VMware HCX Advanced and its Cloud Manager already set up. Instead, you will need to add it as an add-on through the Azure portal. You will still need to download the HCX Connector OVA and set up the virtual appliance on your vCenter Server on-premises. VMware HCX supports up to 25 site pairings in each edition (Advanced and Enterprise) on-premises to cloud or cloud to cloud. HCX Advanced is the default edition in AVS, but you may request HCX Enterprise by submitting a support request to Microsoft support. The HCX Enterprise service may be disabled or turned off; however, HCX Advanced is included in the node fee.

VMware HCX is an application mobility platform for migrating applications, rebalancing workloads, and ensuring business continuity across data centers and clouds.

The following migration types are supported by HCX:

- **Cold migration** – Offline migration of VMs.

- **VMware HCX bulk migration** – The VMware vSphere replication protocols are used by the HCX bulk migration technique to migrate numerous VMs simultaneously to a destination site. Benefits include the following:

  - This technique is intended to migrate numerous VMs concurrently.

  - A predetermined schedule can be used to make the migration complete.

  - Until failover starts, the VMs continue to run at the source site. A reboot would be the equivalent of a service interruption.

- **VMware HCX vMotion live migration** – Zero-downtime live migration of VMs – limited scale. This method uses the VMware vMotion protocol to move a single VM to a remote site.

- **Cloud to cloud migrations** – Direct migrations between VMware Cloud SDDCs moving workloads from region to region or between cloud providers.

- **VMware HCX replication-assisted vMotion** – Bulk live migrations with zero downtime combining HCX vMotion and bulk migration capabilities (HCX Enterprise features). This method also combines the benefits of VMware HCX bulk migration with VMware HCX vMotion live migration.

Throughout this chapter, I will walk you through the steps needed to deploy and configure HCX for your AVS environment.

The following topics will be covered in this chapter:

- Deploying HCX Advanced using the Azure portal
- Downloading and deploying the VMware HCX Connector OVA
- Activating HCX Advanced using the license key from AVS
- Configuring the on-premises HCX Connector

Prerequisites for deploying HCX Advanced

The following requirements need to be in place before deploying HXC Advanced:

- TCP port 443 and UDP port 4500 need to be allowed connectivity between the on-premises HCX Connector and the HCX connector in AVS.

Deploying HCX Advanced using the Azure portal

The steps to deploy HCX Advanced using the Azure portal are as follows:

1. Go to your AVS deployment in the Azure portal. Under **Manage**, select + **Add-ons**. Then, select **Migration using HCX**:

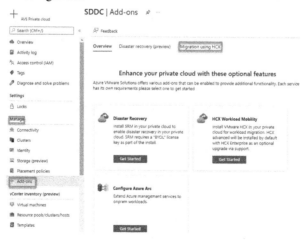

Figure 5.1 – VMware HCX deployment and configuration

2.   Check the **I agree with terms and conditions.** checkbox and then select **Enable and deploy**:

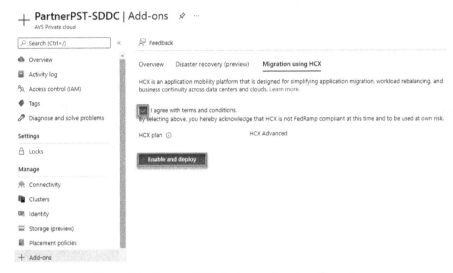

Figure 5.2 – VMware HCX deployment and configuration

3.   Installing HCX Advanced and configuring the Cloud Manager takes around 35 minutes. The HCX Manager URL and HCX keys required for the HCX on-premises connector site pairing will appear on the **Migration using HCX** tab after installation:

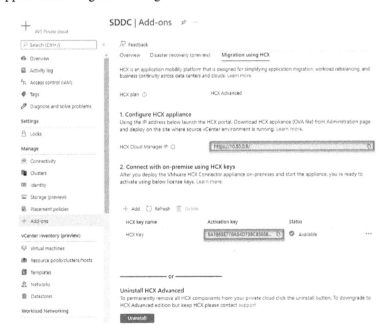

Figure 5.3 – HCX Manager URL and HCX keys

> **Note**
>
> If the HCX key isn't shown after installation, click the **Add** button to produce one, which you may then use for the site pairing.
>
> The HCX Manager URL will always be the .9 parameter from the /22 network that was used to deploy the AVS management components.

Downloading and deploying the VMware HCX Connector OVA

In this section, I will walk you through the downloading and deployment of the HCX Connector OVA file.

Prerequisites

The following ports need to be open from your on-premises vCenter for the IP address `https://x.x.x.9`:

- TCP 443
- UDP 4500

Downloading the HCX Connector OVA file

The steps are as follows:

1. Open a browser window on a computer that has access to your AVS environment on `https://x.x.x.9` on port 443.

2. You will use the vSphere credentials in AVS to log in to the VMware HCX Manager. For the username, enter `cloudadmin@vsphere.local` and then enter the cloud admin password and click **LOG IN**:

Figure 5.4 – HCX Manager login

3.   Under **Administration**, click on **System Updates | Request Download Link**. You may need to wait a few seconds to generate the link if the box is grayed out. Then you get the following screen:

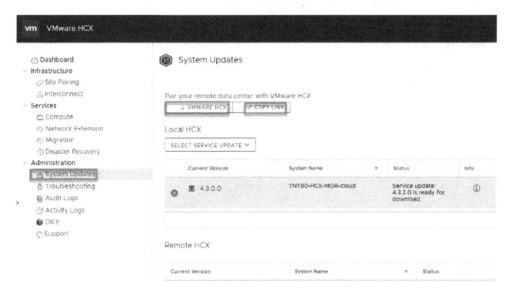

Figure 5.5 – HCX OVA file download options

You have the option to download the file or copy the link to the OVA file.

Deploying the VMware HCX Connector OVA

The steps are as follows:

1. Log into your on-premises vCenter and select the **Hosts and Clusters** icon. Then, click your data center name and from the **ACTIONS** drop-down menu, select **Deploy OVF Template…**:

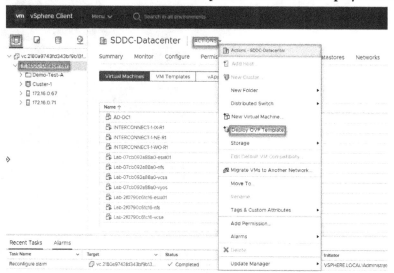

Figure 5.6 – Deploy HCX OVF file

2. Navigate to the location where you saved the OVA file and select **Open**:

Figure 5.7 – HCX OVF file selection

3. Select a name for the VM and select a resource or cluster where you will be deploying the VMware HCX Connector. Then, review the details and required resources and select **NEXT**.

4. Review the license terms, select the required storage and network, and then select **NEXT**.

5. Select the VMware HCX management network segment that you defined during the planning stage. Then, select **NEXT**.

6. In the **Customize template** section, enter all required information, and then select **NEXT**.

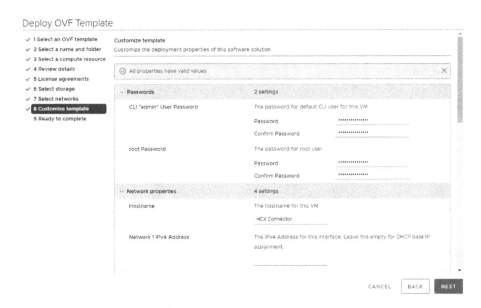

Figure 5.8 – HCX OVF template deployment

7. Verify, and then select **FINISH** to deploy the VMware HCX Connector OVA.

> **Note**
> You will need to manually turn on the virtual appliance. Wait for 10-15 minutes before proceeding.

Activating VMware HCX

First, you must meet the following prerequisites:

- Before you begin, the OVA deployment must be completed. Allow up to 15 minutes for the services to be initialized when the HCX Connector VM is launched.

- Configure firewall rules both on-premises and in AVS to allow TCP port 9443 inbound and outbound.

Now that you have installed the HCX appliance on-premises, you will need to activate it:

1. In the AVS window, select **Manage** | **Add-ons** | **Migrate using HCX**. Copy the activation key.

HCX key name	Activation key	Status
Lab Key	B970631372D04C05BA52DD8... 🗋	✅ Available

Figure 5.9 – HCX Manager activation key

2. Browse to the IP address of the VM that was deployed from the OVA file on port `9443` (`https://HCXManagerIP:9443`) and log in with the following credentials:

    - Username: `admin`

    - Password: Use the vCenter admin password:

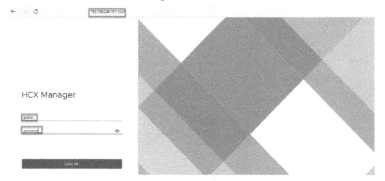

Figure 5.10 – HCX Manager login information

3. Once you are logged into the HCX Manager portal, enter the copied HCX Advanced key, and select **ACTIVATE**. This process can take several minutes to complete.

Figure 5.11 – HCX Manager activation portal

4. In the **Datacenter Location** section, provide the nearest location for installing the VMware HCX Manager on-premises. Then, select **CONTINUE**.

5. In the **System Name** section, modify the name or accept the default and select **CONTINUE**.

6. Select **Yes, Continue**.

7. In **Configure SSO/PSC**, provide the FQDN or IP address of your Platform Services Controller instance, and select **Continue**.

8. Verify that the provided information is correct, and select **Restart**.

Figure 5.12 – HCX Manager System Name

You will see that you have successfully activated your HCX:

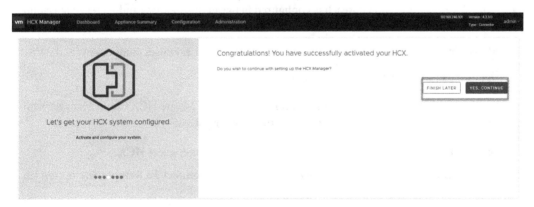

Figure 5.13 – Successful HCX Manager activation portal

As you can see from the preceding screenshot, you have the option to continue with the HCX Connector configuration or you can finish the setup at a later time. We will continue with the configuration in the next section.

Configuring the on-premises HCX Connector

Now that you have installed and activated the HCX add-on, you are now ready for the configuration of the on-premises VMware HCX connector, which will enable the on-premises environment to connect to AVS.

You will learn to do the following:

1. Pair the on-premises VMware HCX Connector with your AVS HCX Cloud Manager
2. Configure the network profile, compute profile, and Service Mesh
3. Check the appliance status and validate that migration is possible when the configuration is completed

When you have completed all the preceding steps, a production-ready environment will be available for you to start creating and migrating your VM.

Adding a site pairing

The prerequisites are as follows:

*   VMware HCX Connector has been installed.
*   Make sure the VMware HCX Enterprise add-on has been enabled through a support request if you intend to use VMware HCX Enterprise. On AVS, VMware HCX Enterprise edition is available and supported at no extra cost.
*   Azure ExpressRoute Global Reach is configured between on-premises and AVS private cloud ExpressRoute circuits.
*   All required ports are open for communication between on-premises components and AVS.
*   Define VMware HCX network segments.

A Site Pair is needed for you to connect or pair your on-premises VMware HCX Cloud Manager with your AVS VMware HCX Connector. Proceed with the following steps:

1. Sign in to your vCenter server on-premises, and under **Home**, select **HCX**.
2. Under **Infrastructure**, select **Site Pairing** and select the **Connect To Remote Site** option (in the middle of the screen).
3. Enter the AVS HCX Cloud Manager URL or IP address that you noted earlier (`https://x.x.x.9`) and the credentials for the CloudAdmin role in AVS. Then, select **Connect**.

Once the connection is successfully made, you should have a screen that looks as follows:

Figure 5.14 – Successful HCX site pairing

Creating network profiles

The VMware HCX Connector automates the deployment of a subset of virtual appliances that need multiple IP segments. Utilize the IP segments you defined during the planning process to generate your network profiles. You'll create four different network profiles:

- Replication
- vMotion
- Management
- Uplink

Let's get started:

1. In your HCX Manager portal, under **Infrastructure**, select **Interconnect | Multi-Site Service Mesh | Network Profiles | CREATE NETWORK PROFILE**:

Figure 5.15 – Creating a network profile

2. For each network profile, select the network and port group, provide a name, and create the segment's IP pool. Then, select **CREATE**.

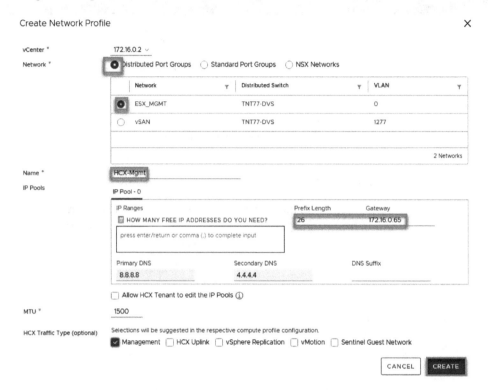

Figure 5.16 – Creating a network profile

After the network profiles are created, they will show up in the **Interconnect** portal.

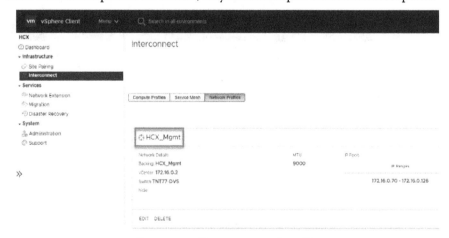

Figure 5.17 – The new network profile

Creating a compute profile

A compute profile contains the compute, storage, and network settings that HCX uses on this site to deploy the interconnected dedicated virtual appliances when a Service Mesh is added:

1. While still logged into the HCX Manager portal, click on **Infrastructure | Interconnect | Compute Profiles | CREATE COMPUTE PROFILE**:

Figure 5.18 – Creating a compute profile

2. Enter a name for the new compute profile and click **CONTINUE**:

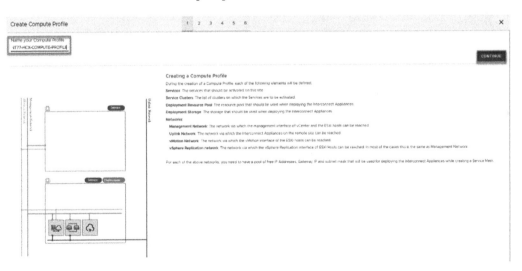

Figure 5.19 – Creating a compute profile name

3. In the **Select Services to be activated** tab, select one or more of the available services to be enabled, and click **CONTINUE**. (Some services cannot be enabled at this time, since you have installed HCX Advanced. HCX Enterprise is required for the additional services to be enabled.)

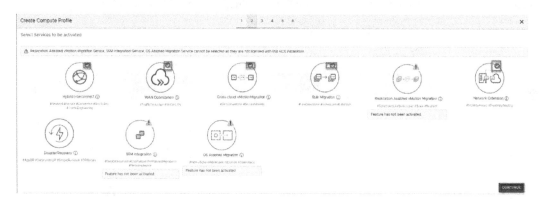

Figure 5.20 – Compute profile services

4. Click on the drop-down icon beside **Select Resources** to make sure you can see your on-premises data center, then select **CONTINUE**.

5. Click on **Select Datastore**, then select the datastore storage resource for deploying the VMware HCX Interconnect appliances. Then, select **CONTINUE**:

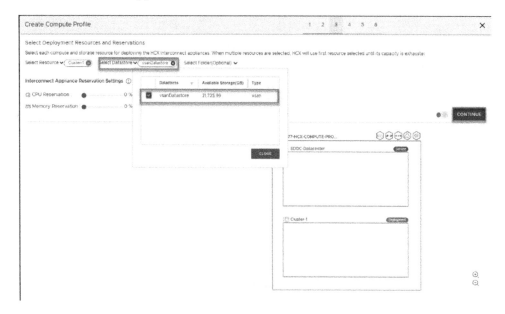

Figure 5.21 – Datastore selection

6. From the **Select Management Network Profile** dropdown, select the management network profile that you created in the previous steps. Then, select **CONTINUE**:

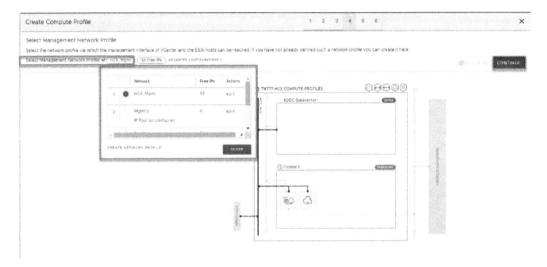

Figure 5.22 – Management network profile selection

7. From the **Select Uplink Network Profile** dropdown, select the uplink network profile you created in the previous procedure. Then, select **CONTINUE**:

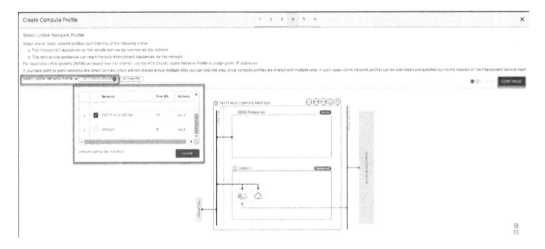

Figure 5.23 – Uplink network profile selection

8. From the **Select vMotion Network Profile** dropdown, select the vMotion network profile that you created in the previous steps. Then, select **CONTINUE**:

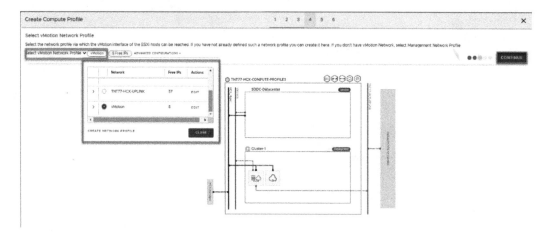

Figure 5.24 – vMotion network profile selection

9. From the **Select vSphere Replication Network Profile** dropdown, select the replication network profile that you created in the previous steps. Then, select **CONTINUE**:

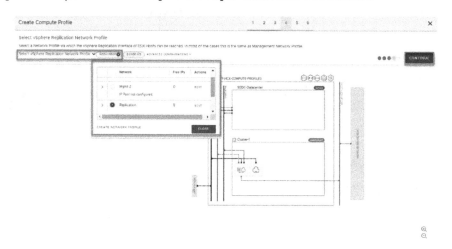

Figure 5.25 – vSphere Replication network profile selection

10. From **Select Distributed Switches for Network Extensions**, select the switches containing the VMs to be migrated to AVS on a **layer-2 (L2)** extended network. Then, select **CONTINUE**. (If you do not plan to migrate any VMs on an L2 extended network, you can skip this step. An L2 extension is needed if you plan to retain the IP addresses of migrated VMs.)

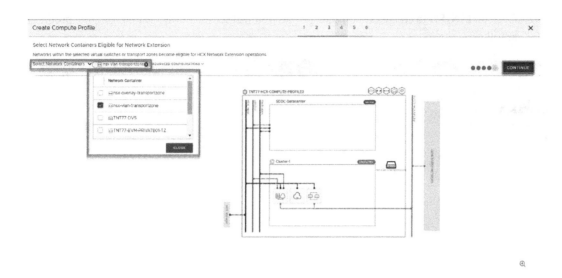

Figure 5.26 – Network extension switch selection

11. Review the connection rules and then select **CONTINUE**:

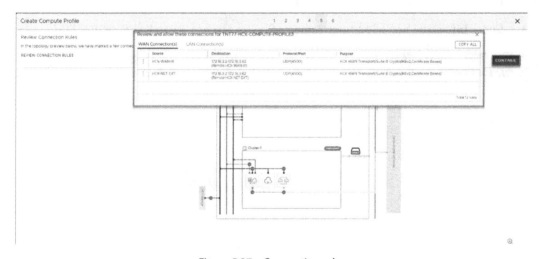

Figure 5.27 – Connection rules

12. Click on **FINISH** to create the compute profile. You will see the newly created compute profile as follows:

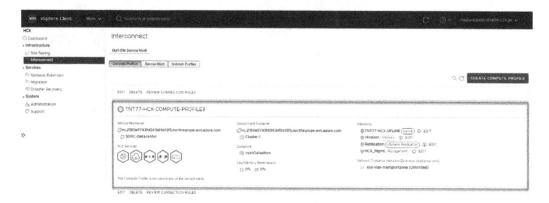

Figure 5.28 – Newly created compute profile

Creating a Service Mesh

The steps to create a Service Mesh are as follows:

1.   Under **Infrastructure**, select **Interconnect** | **Service Mesh** | **CREATE SERVICE MESH**:

Figure 5.29 – Creating a Service Mesh

2. Select the source and remote compute profiles (as shown in the following screenshots) from the drop-down lists, and then select **CONTINUE**. (This option defines where VMs can be migrated to.)

Figure 5.30 – Service Mesh source compute profile

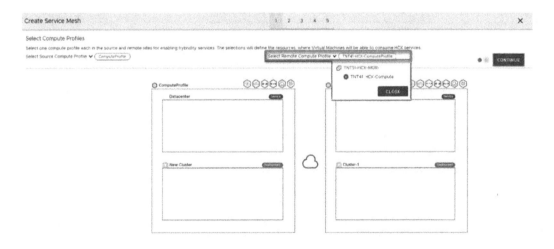

Figure 5.31 – Service Mesh destination compute profile

3. Under **Advanced Configuration - Override Uplink Network profiles**, select **CONTINUE**. The uplink network profiles are used to connect to the network through which the remote site's interconnect appliance can be reached.

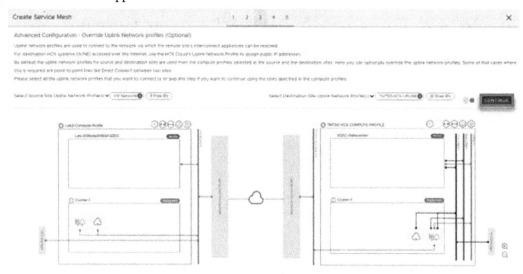

Figure 5.32 – Overriding uplink network profiles

4. On the **Advanced Configuration – Traffic Engineering** page, review and select **CONTINUE**:

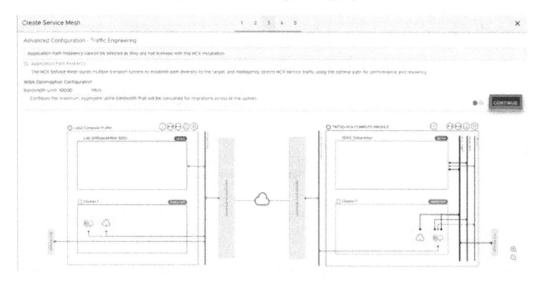

Figure 5.33 – Advance Configuration – Traffic Engineering

5. Now, review the topology preview and select **CONTINUE**. Enter a user-friendly name for this Service Mesh and select **FINISH** to complete:

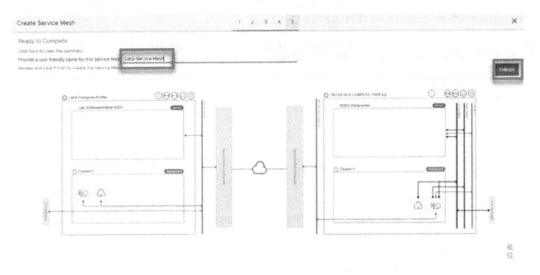

Figure 5.34 – Naming the Service Mesh

The process takes approximately 10-15 minutes to be completed. Once the deployment is completed successfully, all of your services will be green:

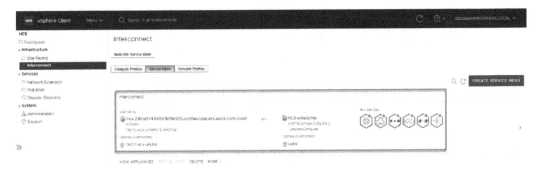

Figure 5.35 – Service Mesh successfully deployed

Validating the Service Mesh appliance status

The status of the HCX connection tunnel should be **Up** and in green. To validate the appliances, select **Interconnect | Service Mesh | VIEW APPLIANCES**:

Figure 5.36 – Viewing Service Mesh appliances status

As you will now see, the tunnels are all green and **Up**:

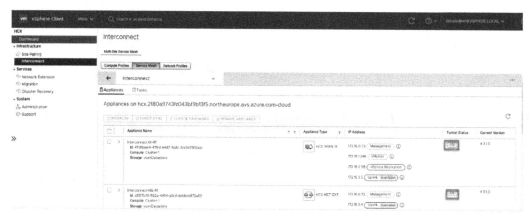

Figure 5.37 – Viewing Service Mesh appliances status

Now that the HCX Tunnels are up, you're ready to use VMware HCX to transfer and safeguard VM workloads from on-premises to AVS. Workload migrations are supported by AVS (with or without a network extension).

Summary

Throughout this chapter, we focused on configuring and deploying HCX in AVS and in the on-premises VMware environment.

VMware HCX is an application mobility technology that makes application transfers, workload rebalancing, and disaster recovery easier between data centers and clouds.

VMware HCX lets you move applications and infrastructure between your on-premises data center and AVS safely and efficiently. HCX provides high-performance, secure, and optimized multi-site connectivity to achieve infrastructure hybridity, and offers multiple options for bidirectional VM mobility with technologies that make it easier to migrate your VM workload to AVS:

- Deploying HCX Advanced using the Azure portal
- Downloading and deploying the VMware HCX Connector OVA
- Activating HCX Advanced using the license key from AVS
- Configuring the on-premises HCX Connector

As you have learned, VMware HCX plays an integral part in your on-premises to AVS migration strategy. While there are other methods of migrating a VM workload from on-premises to AVS, VMware HCX simplifies the process while remaining a secure and high-performance solution.

In the next chapter, we will be focusing on NSX-T Manager, which serves as the networking component for AVS.

6
Networking in AVS using NSX-T

By default, AVS includes NSX-T Data Center. AVS is preconfigured with an NSX-T Data Center Tier-0 gateway operating in active-active mode and a default NSX-T Data Center Tier-1 gateway in active-standby mode. These gateways allow you to join the segments (logical switches) and make connections in the East-West and North-South directions.

After installing AVS, you can use the Azure portal to set up the required NSX-T Data Center objects. It is designed for users unfamiliar with NSX-T Manager and provides a simplified view of the NSX-T Data Center activities that a VMware administrator needs every day.

Getting familiar with the NSX-T functionalities is vital, as this will be your default networking tool in AVS. Customers do have the option to use a **network virtual appliance** (**NVA**) as the networking tool for AVS. NSX-T is deployed with the following services and functionalities by default:

- Segments
- DHCP
- DNS
- Port mirroring

The following diagram illustrates an NSX-T high-level architecture:

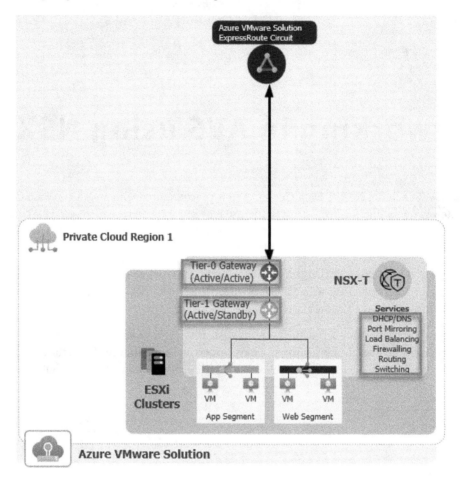

Figure 6.1 – NSX-T high-level architecture

After a successful AVS private cloud deployment, you'll need to configure the NSX-T network segments using NSX-T Manager or the Azure portal. The segments are logical switches that your AVS workloads require.

These segments are visible in AVS (in the Azure portal), NSX-T Manager, and vCenter Server once configured. By default, NSX-T includes an NSX-T Tier-0 gateway in active-vCenter active mode and a default NSX-T Tier-1 gateway in active-standby mode.

Virtual machines (**VMs**) will not have IP addresses until statically or dynamically assigned from a DHCP server or DHCP relay.

Throughout this chapter, we will focus on the following topics:

- Configuring DHCP for AVS
- Adding an NSX-T segment using the Azure portal
- Adding an NSX-T segment using NSX-T Manager
- Verifying the newly created network segment
- Configuring DNS for AVS
- Deploying a test VM and connecting it to the newly created segment

Configuring DHCP for AVS

In this section, I will walk you through the steps to create a DHCP server using the Azure portal.

In a private cloud environment, applications and workloads need name resolution (DNS) and DHCP services for IP address allocation. You'll need a good DHCP and DNS infrastructure to deliver these services. You can construct a VM to provide these services in a private cloud environment.

It is recommended that, instead of forwarding broadcast DHCP traffic across the WAN back to on-premises, you utilize the DHCP service integrated into NSX or a local DHCP server in the private cloud.

A DHCP server or DHCP relay is needed before you can create an NSX-T segment if you plan to use DHCP in AVS.

Prerequisites

You will need an AVS infrastructure with access to the NSX-T Manager and the vCenter Server interfaces.

Using the Azure portal to create a DHCP server or relay for AVS

From the Azure portal, you can create a DHCP server or relay. These services connect to the Tier-1 gateway that was created when you deployed the AVS infrastructure. You will need to specify the DHCP ranges for the segments that need to utilize the DHCP service:

1. Log in to your Azure portal, select your AVS private cloud, and then, under **Workload Networking**, select **DHCP | Add**.

2. Select **DHCP Server** or **DHCP Relay**. Provide a name for the server and an IP address for the server, and then click **OK**:

Figure 6.2 – Creating a DHCP server in the Azure portal

If you are creating a DHCP relay, you will need to provide a name for the relay server and provide an IP address, and then click **OK**:

Figure 6.3 – Creating a DHCP relay in the Azure portal

The steps for creating a DHCP server or relay for AVS are as follows:

1. Log in to your **NSX-T Manager** portal. If this is the first time you are logging in, accept the **End User License Agreement** (**EULA**) and click **CONTINUE**. Click **SAVE** on the **Customer Experience Improvement Program** window.

2. Select **Networking | DHCP**, and then select **Add DHCP PROFILE**.

3. Enter a name for your DHCP server.

4. Under **Profile Type**, select **DHCP Server**. Enter an IP address in the form of x.x.x.x/x under **Server IP Addresses**.

5. Enter a time-out in seconds under **Lease Time (seconds)**, and then click **SAVE**.

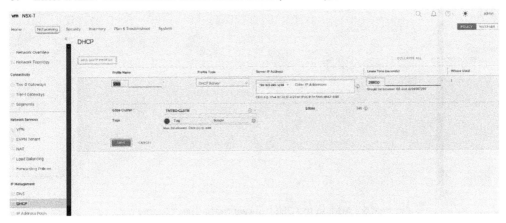

Figure 6.4 – Creating a DHCP Server in NSX-T Manager

You will now see a new **DHCP Server** in the list of servers.

Figure 6.5 – Newly created DHCP Server in NSX-T Manager

Now, we need to make sure that the Tier-1 gateways can access the newly created DHCP Server for IP distribution to network segments that will be created later.

6. From the **NSX-T Manager** portal, click on **Tier-1 Gateways**. Select the vertical ellipsis (**…**) on the Tier-1 gateway and then select **Edit**.

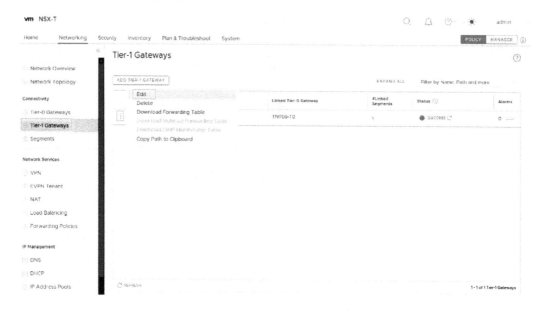

Figure 6.6 – Adding the DHCP Server to the Tier-1 Gateways

7. Select **Set DHCP Configuration**.

Figure 6.7 – Set DHCP Configuration

8. Select **DHCP Server** from the drop-down options under the **Set DHCP Configuration** tab. Select the DHCP Server that you created earlier and click **SAVE**.

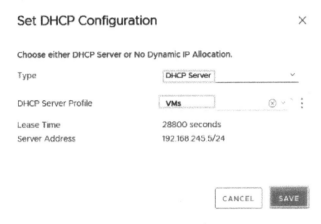

Figure 6.8 – Set DHCP Configuration

9. Click on **SAVE** again. You will now notice that the DHCP option is now **Local | 1 Servers**.

Figure 6.9 – Tier-1 Gateways screen updated with the DHCP server

You will now see the DHCP server information in the Azure portal.

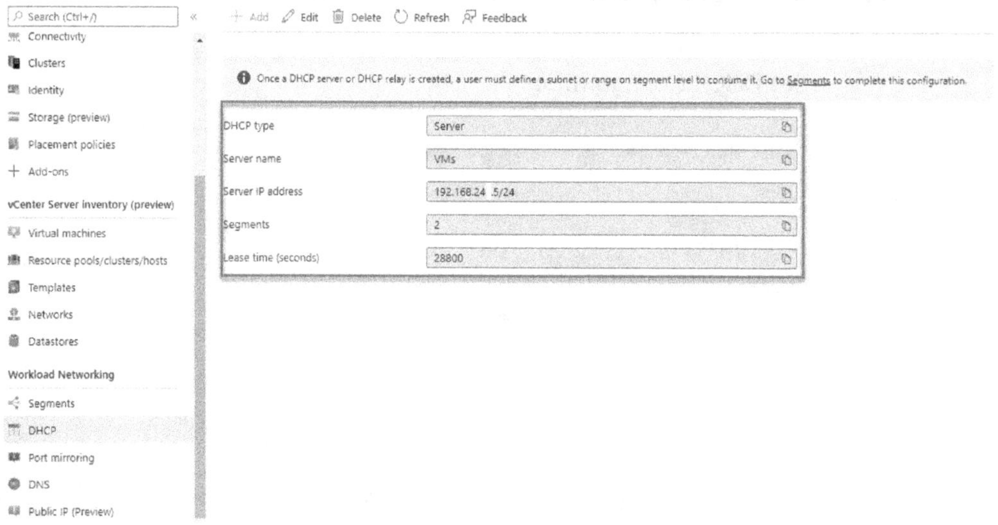

Figure 6.10 – DHCP server information in the Azure portal

With that, you have learned how to create a DHCP server and a DHCP server relay in AVS. As you have seen, it is very straightforward to configure DHCP services in AVS.

In the next section, you will learn how to create segments in AVS using both the Azure AVS portal and the NSX-T management portal.

Adding an NSX-T segment using the Azure portal

I will now walk you through the steps to create an NSX-T segment using the Azure portal:

1. From the Azure portal, navigate to your AVS private cloud, and, under **Workload Networking**, click on **Segments | Add**.

2. Provide the required details for the new segment and click **OK**:

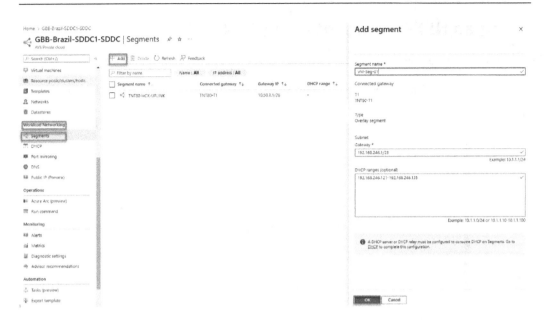

Figure 6.11 – Adding an NSX-T segment in the Azure portal

The following are the details you have to add:

- **Segment name** – This is the name of the segment that will be visible in vCenter.

- **Subnet gateway** – This is the gateway IP address for the new segment. You will need to enter the IP address with the subnet mask ($x.x.x.x/x$). Similarly, on-premises, VMs are attached to a logical segment and all VMs connecting to a segment belong to the same subnet. IP addresses will be issued from this segment to all the VMs that are connected to it.

- **DHCP** – DHCP ranges for a logical segment are optional. You will need to configure a DHCP server or DHCP relay to consume DHCP on segments.

The segment will now be visible in NSX-T Manager and vCenter.

Adding an NSX-T segment using NSX-T Manager

The steps for this are as follows:

1.   From your **NSX-T Manager** portal, click on **Networking | Segments**. Then, click on **Add Segment**.

Figure 6.12 – Adding an NSX-T segment in the NSX-T Manager portal

2.   Name the segment.

3.   Set your Tier-1 gateway (**TNTxx-T1**) as **Connected Gateway**.

4.   Select the **Transport Zone** overlay (**TNTxx-OVERLAY-TZ**).

5.   Enter your network CIDR IP address (x.x.x.x/x), and then click on **SAVE** at the bottom of the screen.

Figure 6.13 – Adding an NSX-T segment in the NSX-T Manager portal

Now, you have learned how to create segments in AVS using both the Azure AVS portal and the NSX-T management tool. It is recommended to use different segments for different workload types or organizational segmentations. You can create additional Tier-1 gateways to further help you segment your different workloads.

In the next section, you will learn how to add new segments to a DHCP server using the NSX-T Manager portal. This is very important, as all your segments should use the same DHCP server that you configured earlier. Each new segment needs to be added to a DHCP server for IP address distribution to new VMs. You can do this during the configuration of a new segment or edit a segment after it is created:

1.   In your **NSX-T Manager** portal, click on **Networking | Segments**, then click on the ellipsis next to the segment you created earlier, and click on **Edit**.

Figure 6.14 – Adding an NSX-T segment to a DHCP server

2.   Click on **SET DHCP CONFIG**.

3.   Click on the toggle switch to enable DHCP for this segment (by default, DHCP is disabled for new segments).

4.   Enter your desired DHCP range and click **APPLY**.

5. Click **SAVE** on the next screen.

Figure 6.15 – Adding an NSX-T segment to a DHCP server

Verifying the newly created network segment

You will now verify the new segment that was just created. **VM-2-Seg-2** was the new segment create:.

1. In the **NSX-T Manager** portal, click on **Networking | Segments**.

Figure 6.16 – New segment verification in NSX-T Manager

2. Log in to your vCenter, and select **Networking | SDDC-Datacenter**.

Figure 6.17 – New segment verification in vCenter

Configuring DNS for AVS

vCenter Server and other AVS administration components can only resolve name records accessible via public DNS by default. Certain hybrid use cases – customer-managed systems such as vCenter Server and Active Directory – need AVS administration components to resolve name records from privately hosted DNS to work effectively.

Through the NSX-T Manager DNS service, you can build conditional forwarding rules for the required domain name to a specified set of private DNS servers using Private DNS for AVS administration components.

You'll get a DNS service and a default DNS zone in the AVS deployment. You must establish an FQDN zone and apply it to the NSX-T Manager DNS Service to allow AVS management components to resolve records from your private DNS systems. DNS requests for each zone are conditionally forwarded by the DNS Service depending on the external DNS servers configured in that zone. The DNS Forwarder Service in NSX-T Manager is used for this feature.

> **Important note**
>
> The DNS Service is associated with up to five FQDN zones. Each FQDN zone is associated with up to three DNS servers.

Configuring a DNS forwarder

The steps are as follows:

1. In the Azure portal, navigate to your AVS private cloud. Under **Workload Networking**, select **DNS | DNS Zone**. Click **Add**.

Figure 6.18 – Creating a new DNS zone

2. Select the **FQDN zone** option under **Type**. Provide the DNS zone name and the FQDN for the domain name, as well as up to three DNS server IP addresses in the format of x.x.x.x. Select **OK**.

Figure 6.19 – Configuring a new DNS zone

It will take several minutes to complete. You will see a message that the DNS zone was created in Azure Notifications.

Important note

While NSX-T Manager may be used to execute certain operations in your private cloud, all configuration modifications to the default Tier-1 gateway must be done by editing the DNS service from the Simplified Networking experience in the Azure portal for private clouds built on or after July 1, 2021.

---

3.   Click on the **DNS service** option and then click **Edit**.

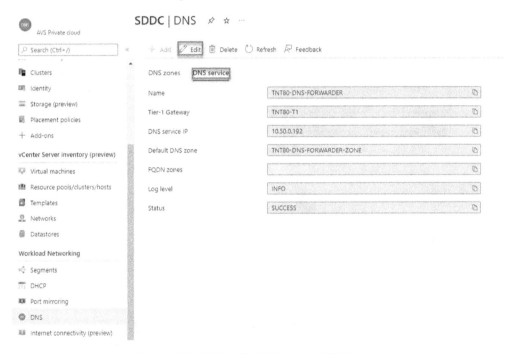

Figure 6.20 – Editing the DNS service FQDN zones

4. From the **FQDN zones** drop-down selection, select the new FQDN and then click **OK**.

Figure 6.21 – Editing the DNS service FQDN zones (part 2)

It takes a few minutes to complete, and when it's done, you will receive the **Completed** notification under Azure Notifications. At this point, AVS management components should be able to resolve DNS entries from the FQDN zone supplied to the NSX-T Manager DNS Service.

DNS name resolution verification for AVS

You have a few choices for verifying name resolution activities once you've set up the DNS forwarder:

1. Log in to the **NSX-T Manager** portal, select **Network | DNS**, and expand the DNS Forwarder Service.

Figure 6.22 – DNS resolution verification

2. Select **View Statistics**, and from the **Zone Statistics** dropdown, select the FQDN zone that you created earlier.

Figure 6.23 – DNS resolution statistics

The top half of the page displays data for the entire service, while the bottom half displays information for the zone you have selected. The queries forwarded to the DNS services that were defined when the FQDN zone was set can be seen in this example.

Deploying a test VM and connecting it to the newly created segment

Now that a new segment has been created and a DHCP server has been configured for that segment, we can ensure that a VM deployed to that segment will be able to receive an IP address from the DHCP server.

There are multiple ways to create a VM in vCenter. For this exercise, we will create a VM from an imported template:

1. Log in to **vCenter** | **VMs and Templates** | **Templates** | your template (in this case, it is **Windows-2022-Template**) | **ACTIONS** | **New VM from This Template**.

Figure 6.24 – New VM from an OS template

2.   Enter a name for the VM, select a location for the VM (select the data center), and click on **NEXT**.

Figure 6.25 – New VM from an OS template (part 2)

3.   Select **Cluster-1** as the destination compute resource and click **NEXT**.

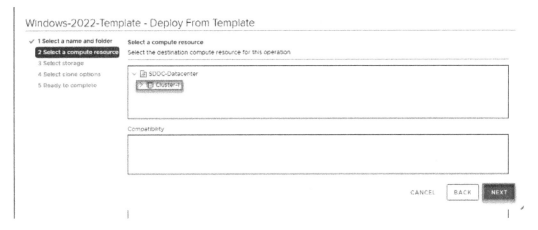

Figure 6.26 – Compute destination for the new VM

4.   Select **vsanDatastore** for the VM storage and click **NEXT**.

Figure 6.27 – Storage option for the new VM

5.   Under the clone options, check the checkbox next to **Power on virtual machine after creation** and click **NEXT**.

Figure 6.28 – Clone options for the new VM

6. Review the options that you have selected and click on **FINISH**.

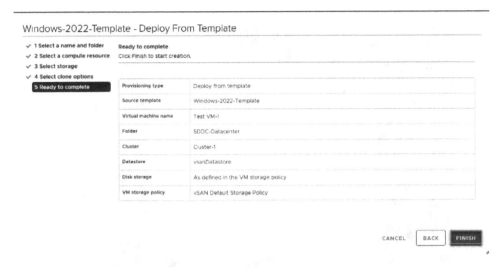

Figure 6.29 – Overview of the new VM

Moving a VM to a different network segment

The VM was created on the **VM-Seg-1** network segment and we will move it to the new network segment that was created previously. To do this, use the following instructions:

1. In your vCenter, click on **VMs and Templates | Test VM-1** (this is the new VM that was created) | **Networks**. Right-click on **VM-Seg-1** and click on **Migrate VMs to Another Network**.

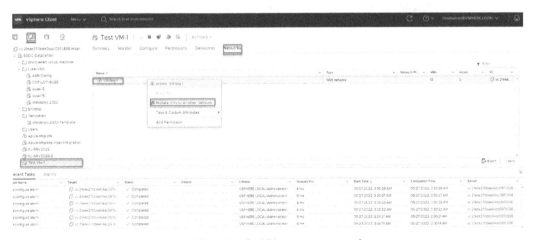

Figure 6.30 – Migrating the VM to a new network segment

2. Click on **BROWSE**, then select **VM-2-Seg-2** (this is the newly created network segment), and click **OK**.

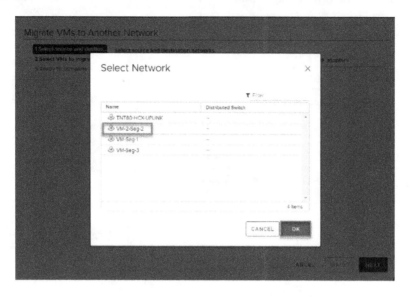

Figure 6.31 – Browsing for a different network segment

3. Click on **NEXT**.

4. Check the checkbox next to **Test VM-1** and click **NEXT**.

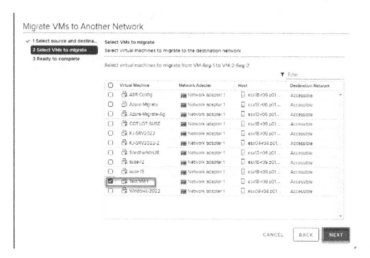

Figure 6.32 – Selecting the VM to be migrated to a different network segment

5. Click on **FINISH**.

6. Click on **Test VM-1**. You will notice that the VM is now on the newly created network segment, **VM-2-Seg-2**.

Figure 6.33 – New VM migrated to the new network segment

7. Click on **Test VM-1 | Summary**. You will notice that the DNS name is from the domain name that was used when the DNS FQDN was created.

The IP address is from the DHCP range configured on the DHCP server.

Figure 6.34 – New VM summary in vCenter

As you have seen throughout this chapter, NSX-T Manager is a powerful yet easy-to-use networking tool that is deployed and configured by default when you deploy AVS in Azure. Customers who would like to use another networking solution can use a **Network Virtual Appliance** (**NVA**) in AVS.

Summary

This chapter covered NSX-T Manager for AVS. When you deploy AVS, NSX-T is the default networking and security management stack.

NSX-T controls the traffic flow between North/South and East/West in AVS. NSX-T is deployed with the following services and functionalities by default:

- Segments
- DHCP
- DNS
- Port mirroring

You have learned how to configure DHCP and add a segment using both the Azure portal and NSX-T Manager. You have also learned how to configure DNS and how to verify that NSX-T Manager is configured correctly for your AVS environment by deploying a new VM and verifying the different parameters.

In the next chapter, I will walk you through creating and configuring a secure vWAN Hub for internet connectivity and traffic inspection for AVS.

Part 3: Configuring Your AVS

This part will cover internet access from AVS, traffic inspection, storage options, SRM for disaster recovery, and the management and governance of the AVS environment.

This part comprises the following chapters:

- *Chapter 7, Creating and Configuring a Secure vWAN Hub for Internet Connectivity*
- *Chapter 8, Inspecting Traffic for AVS*
- *Chapter 9, Adding Additional Storage to the AVS Datastore*
- *Chapter 10, Working with VMware Site Recovery Manager*

7

Creating and Configuring a Secure vWAN Hub for Internet Connectivity

Utilizing AVS in conjunction with the Azure cloud ecosystem necessitates a distinct set of architectural considerations for cloud-native and hybrid situations. This chapter will discuss how to connect to the internet via a secure **virtual WAN (vWAN)**. We will explore the critical factors and best practices for networking and connecting to, from, and inside Azure and AVS deployments.

The following topics will be covered in this chapter:

- Azure vWAN in Azure
- Creating a vWAN in Azure
- Deploying Azure Firewall to secure the vWAN
- Creating an Azure Firewall policy for the AVS internet connection

Azure vWAN in Azure

Azure vWAN is a networking solution that combines a variety of networking, security, and routing capabilities into a single operating interface. The following are some of its essential features:

- VPN ExpressRoute inter-connectivity
- Site-to-site VPN connectivity
- Private connectivity (ExpressRoute)
- Intra-cloud connectivity (transitive connectivity for virtual networks)
- Routing, Azure Firewall, and encryption for private connectivity

- Branch connectivity (via connectivity automation from vWAN partner devices such as SD-WAN or VPN CPE)

- Remote user VPN connectivity (point-to-site)

To get started with vWAN, you don't need to have all these use cases. You may begin with a single use case and then alter your network as your needs change.

In this chapter, we will be focusing on private connectivity using ExpressRoute, routing, and Azure Firewall. We will also look at encryption for private connectivity.

For branches (VPN/SD-WAN devices), users (Azure VPN/OpenVPN/IKEv2 clients), ExpressRoute circuits, and virtual networks, the vWAN design is a hub-and-spoke architecture with scalability and performance automatically built in. It allows a worldwide transit network design, with the cloud-hosted network "hub" enabling a transitive connection between endpoints scattered throughout various spokes.

For this use case, AVS will be the spoke while vWAN will be the hub.

In *Figure 7.1*, Azure Firewall is used in the secured vWAN hub for AVS egress and ingress L4 traffic inspection. An Application Gateway is recommended for L7 load balancing and SSL offloading. The connectivity back to the customers' on-premises can either be an Azure ExpressRoute, a site-to-site VPN, or a software-defined WAN:

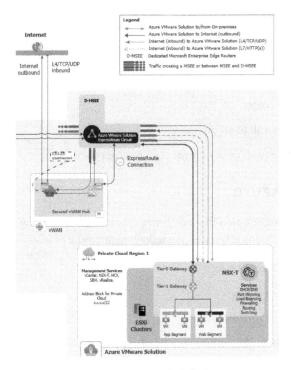

Figure 7.1 – Secured vWAN Hub with default route propagation

Advantages of a vWAN

The following are some of the advantages of a vWAN:

- **Hub-and-spoke networking systems that are integrated**: Connect on-premises sites to an Azure hub by automating site-to-site setup and connection

- **Spoke setup and configuration through automation**: Connect your virtual networks and workloads to the Azure hub simply and securely

- **Intuitive troubleshooting**: Within Azure, you can observe the end-to-end flow and then utilize that knowledge to take the necessary measures

vWAN types

There are two varieties of virtual WANs: Standard and Basic. The following table details the settings available for each class.

Virtual WAN Type	Hub Type	Available Configuration
Basic	Basic	Site-to-site VPN only
Standard	Standard	ExpressRoute User VPN (P2S) VPN (site-to-site) Inter-hub and VNet-to-VNet transiting through the virtual hub Azure Firewall NVA in a virtual WAN

Table 7.1 – vWAN types

As you can see from the preceding table, you will need to configure a Standard vWAN to route AVS traffic to the internet.

Creating and configuring a secured vWAN in Azure includes several different resources and configurations. Throughout the rest of this chapter, you will learn how to securely connect your AVS environment to the internet via a secured vWAN.

You will learn how to do the following:

- Create a vWAN in Azure
- Create a virtual hub and an ExpressRoute gateway in the vWAN
- Connect the AVS ExpressRoute circuit to the hub gateway
- Change the size of the gateway
- Add Azure Firewall to secure the vWAN
- Create an Azure Firewall policy for the AVS internet connection
- Add a rule to the Firewall policy
- Associate the Firewall policy with the hub
- Route the AVS traffic to the vWAN hub

Creating a vWAN in Azure

Azure vWAN is a networking solution that combines a variety of networking, security, and routing capabilities into a single operating interface.

Prerequisites

The prerequisites for creating a vWAN in Azure are as follows:

- An AVS environment that you will be connecting to.
- You must have a public-facing IP address terminating on an on-premises VPN device.
- vWAN creates and uses a hub, which is a virtual network. Get an IP address range for the Azure region where you will be deploying the hub. The address range for the hub must not overlap with any of the virtual networks to which you are connected. It also can't be in the same address range as your on-premises or AVS address ranges.

Follow these steps to create a vWAN:

1. Log in to the Azure portal and, in the **Search resources** bar, type Virtual WAN and press *Enter*.
2. Click on **Virtual WANs** from the displayed results. On the **Virtual WANs** page, click + **Create**. This will open the **Create WAN** page.

3. On the **Create WAN** page, fill in the fields on the **Basics** tab. See the example shown in the following screenshot:

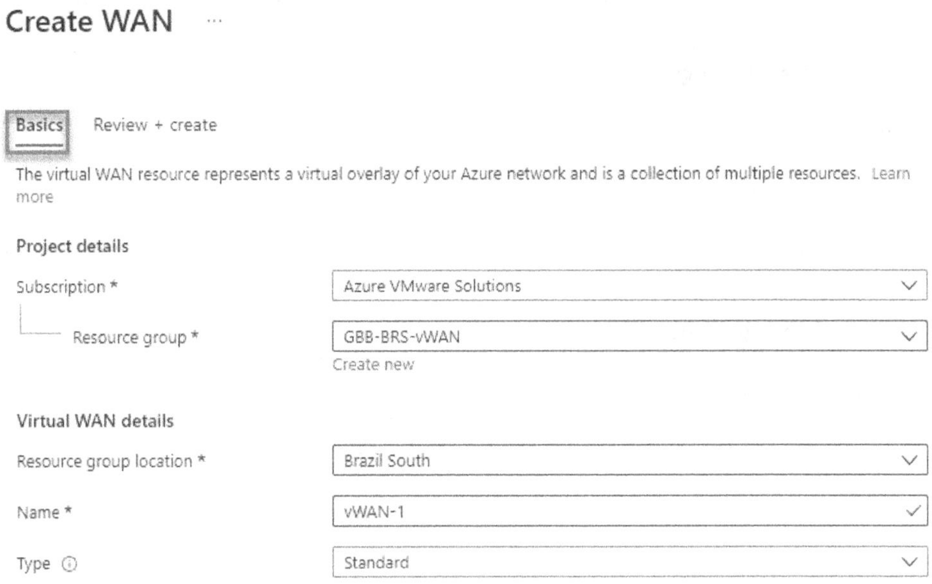

Create WAN   ...

Review + create

The virtual WAN resource represents a virtual overlay of your Azure network and is a collection of multiple resources. Learn more

Project details

Subscription *	Azure VMware Solutions
Resource group *	GBB-BRS-vWAN

Create new

Virtual WAN details

Resource group location *	Brazil South
Name *	vWAN-1
Type ⓘ	Standard

Figure 7.2 – Creating a vWAN

The following are more details of the fields:

- **Subscription**: Select the subscription that you plan on using for the vWAN.

- **Resource group**: You can use an existing resource group or create a new one.

- **Resource group location**: Choose an Azure region from the dropdown. A WAN is a global resource not limited to a single region. However, you must first choose a region to manage and find the WAN resource you created.

- **Name**: Type in a name for your vWAN.

- **Type**: There are two different types of vWAN – Standard and Basic. Since we will be connecting to the ExpressRoute circuit from AVS, you will need to select Standard. A Basic WAN can only be connected to a VPN connection.

4. After filling out the fields, select **Review + Create** at the bottom of the page.

5. After validation passes, click **Create** to create the vWAN.

This process only takes a few minutes to complete.

Creating a virtual hub and an ExpressRoute Gateway in the vWAN

A virtual hub is a virtual network that vWAN creates and uses. It's the core of your vWAN network in a region. It may include VPN and ExpressRoute gateways. You can either create the gateway while creating a new virtual hub or create the gateway in an existing virtual hub using the edit feature. You will create an ExpressRoute gateway for your virtual hub in this section.

ExpressRoute gateways are deployed in 2 Gbps increments. One scale unit equals 2 Gbps, with a maximum of 10 scale units equaling 20 Gbps. It takes roughly 30 minutes for the virtual hub and gateway to be deployed successfully.

Please note that once the hub is created, you will be accruing charges for it, even if you have not connected it to any sites.

To create a virtual hub, follow these steps:

1. Go to the vWAN that you created earlier. Under **Connectivity**, click on **Hubs**:

Figure 7.3 – Creating a hub in a vWAN

2. On the **Hubs** page, select **+New Hubs**. This will open the **Create virtual hub** page:

> **Important note**
>
> The hub will be created in the same subscription and resource group where the vWAN is created. You cannot change this option.

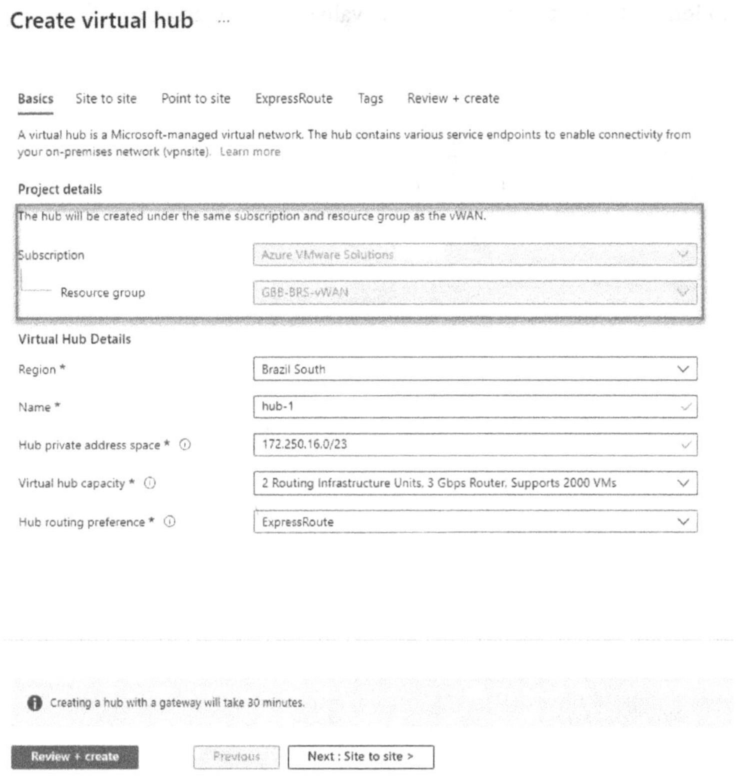

Create virtual hub ...

Basics   Site to site   Point to site   ExpressRoute   Tags   Review + create

A virtual hub is a Microsoft-managed virtual network. The hub contains various service endpoints to enable connectivity from your on-premises network (vpnsite). Learn more

Project details

The hub will be created under the same subscription and resource group as the vWAN.

| Subscription | Azure VMware Solutions |
| Resource group | GBB-BRS-vWAN |

Virtual Hub Details

Region *	Brazil South
Name *	hub-1
Hub private address space * ⓘ	172.250.16.0/23
Virtual hub capacity * ⓘ	2 Routing Infrastructure Units, 3 Gbps Router, Supports 2000 VMs
Hub routing preference * ⓘ	ExpressRoute

ⓘ Creating a hub with a gateway will take 30 minutes.

Review + create   Previous   Next : Site to site >

Figure 7.4 – Hub subscription and resource group

3. Complete the following fields on the **Basic** tab:

- **Region**: The region in which you want to create your virtual hub

- **Name**: The name you choose for the virtual hub

- **Hub private address space**: The hub's address range in CIDR notation. The minimum address space is /24 to create a hub

- **Virtual hub capacity**: Select from the dropdown

- **Hub routing preference**: This field is only available as part of the virtual hub routing preference preview and can only be viewed in the preview portal

4. Click on the **ExpressRoute** tab and click on **Yes** to agree to create an ExpressRoute gateway.

5.   Select an option for the **Gateway Scale Units** value from the dropdown list:

Figure 7.5 – ExpressRoute sections

6.   Click on **Review + Create** for the selections to be validated:

Figure 7.6 – Virtual hub validation

7. Click on **Create**. This process takes up to 30 minutes to complete.

8. Once the virtual hub has been successfully deployed, you can view the details of the ExpressRoute gateway that was created during the hub deployment. To see the details, go to the hub and click on **ExpressRoute**:

Figure 7.7 – ExpressRoute gateway details

Connecting the AVS ExpressRoute circuit to the hub gateway

ExpressRoute Standard or Premium circuits in locations that support ExpressRoute Global Reach can connect to a vWAN ExpressRoute gateway and use all the vWAN transit features (VPN-to-VPN, VPN, and ExpressRoute transit). Now that the gateway has been set up, the AVS ExpressRoute circuit can connect. ExpressRoute Standard and Premium circuits in locations outside of Global Reach can connect to Azure resources, but they can't use vWAN transit features. vWAN hubs can also be used with ExpressRoute Local.

To connect the ExpressRoute circuit to the gateway, follow these steps:

1. Log in to the Azure portal and go to your AVS deployment.

2. Under **Manage**, click on **Connectivity | ExpressRoute**.

3.   Click on + **Request an authorization key**. Give the key a name and click **Create**:

Figure 7.8 – Requesting an ExpressRoute authorization key

4.   When the authorization key is created, copy it to your clipboard. Also, copy the ExpressRoute ID. You will need that information to connect to the hub gateway.

5.   Locate the hub that you created and click on **ExpressRoute | + Redeem authorization key**.

6.   Paste the authorization key that you created earlier into the **Authorization Key** section.

7.   Paste the ExpressRoute ID into the **Peer circuit URI** section. Click on the check box next to **Automatically associate this ExpressRoute circuit with the hub**. Click **Add**:

Figure 7.9 – Redeeming an ExpressRoute authorization key

It takes about 10 to 15 minutes for the ExpressRoute circuit to be connected to the hub gateway.

Once the AVS ExpressRoute circuit is connected to the hub gateway, it will appear on the hub page under **Manage ExpressRoute circuits**:

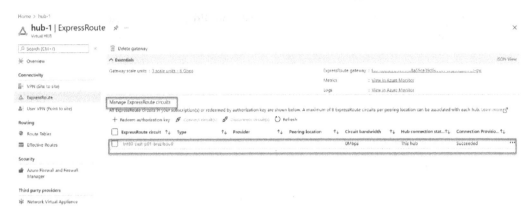

Figure 7.10 – Connected ExpressRoute circuit

You can also see that an ExpressRoute circuit has been connected to the hub from the vWAN:

Figure 7.11 – Viewing a connected ExpressRoute circuit status from the vWAN

Changing the size of the gateway

You may need to change the size of the gateway at some point. You may need to do so to increase (scale up) or decrease (scale down) the throughput.

To do this, follow these steps:

1. Connect to the hub and, under **Connectivity**, click on **ExpressRoute**.

2. Click on the current scale unit next to **Gateway scale units**.

3. Select the new scale unit from the drop-down list from **Edit ExpressRoute Gateway**, then click on **Edit**:

Figure 7.12 – Changing the size of a gateway

> **Important note**
>
> Updating any components to the hub can take up to 30 minutes to complete.

Deploying Azure Firewall to secure the vWAN

Now that we have a vWAN and a hub, we need to secure it by deploying Azure Firewall. In addition to securing the vWAN, Azure Firewall will also be used as the default route for all internet traffic from AVS. It will also be used for traffic inspection.

Prerequisites

The prerequisites for deploying Azure Firewall are:

*   A deployed vWAN
*   A deployed hub in the vWAN

Deploying Azure Firewall

The steps are as follows:

1. On the **Overview** page of your deployed virtual WAN, click on the hub that you will be converting into a secure hub:

Figure 7.13 – Selecting the hub to be converted

2. On the virtual hub page, you will see that you have two options to deploy Azure Firewall to the hub. In this example, we will select the **Azure Firewall** option:

Figure 7.14 – Options to deploy Azure Firewall to the hub

3. After you make your selection, you will be presented with the option to select the virtual hub that you will be converting into a secure hub.

4. Select the hub under **Select virtual hubs** and click **Next: Azure Firewall**:

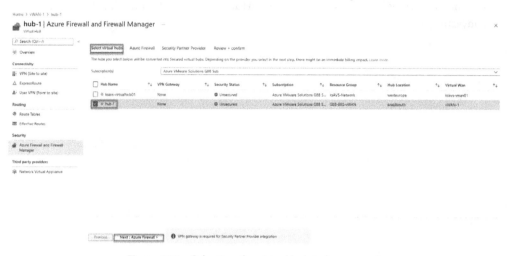

Figure 7.15 – Selecting the virtual hub to be secured

5. From the **Azure Firewall** tab, make sure **Azure Firewall** is **Enabled**. Select the desired Azure Firewall tier (**Standard** or **Premium**).

6. Specify the number of public IP addresses you need for the firewall.

7. The subscription will default to the subscription where the vWAN is deployed to.

8. Select **Default Deny Policy**. This policy blocks everything by default, so you will need to create the policy that you need afterward.

9. Click **Next: Security Partner Provider**:

Figure 7.16 – Azure Firewall option

10. Make sure that the **Security Partner Provider** option is set to **Disabled**. You won't be needing that for this exercise. Click on **Next: Review + confirm**.

11. After you see the **Validation passed** confirmation, click on **Confirm**:

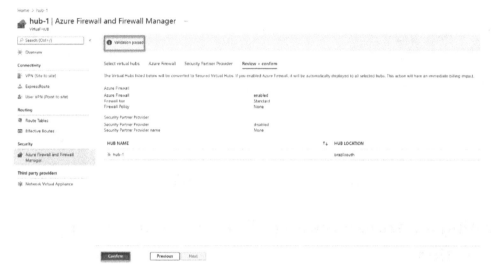

Figure 7.17 – Azure Firewall validation passed

This process can take up to 10 minutes to complete.

12. After the hub has been converted into a secure hub, you can view it on the vWAN **Overview** page. You will notice that the **Azure Firewall** entry in the table now reads **Deployed**:

Figure 7.18 – Azure Firewall visible in the vWAN

13. Click on the hub; you will see that Azure Firewall has a green checkbox and reads **Secured**. The vWAN is now secured:

Figure 7.19 – vWAN is now secured by Azure Firewall

Creating an Azure Firewall policy for the AVS internet connection

A firewall policy is a set of rules that specify how traffic is sent to one or more secured virtual hubs. You will now create an empty firewall policy without any rules using the Azure Firewall Manager. You will add the rules later:

1. From the Azure portal, type `firewall manager` in the **resources, services and docs** search bar and hit *Enter* on your keyboard.

2. On the **Firewall Manager** page, select **Azure Firewall Policies** under the **Security** section. Click on + **Create Azure Firewall Policy**:

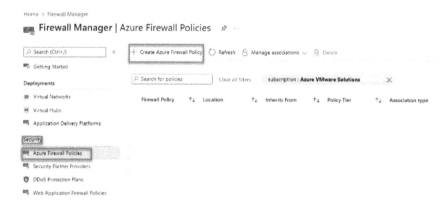

Figure 7.20 – Creating an Azure Firewall policy

3. On the **Create an Azure Firewall Policy** page, make sure you have the correct subscription selected. Create a new resource group or use an existing one.

4. Under **Policy details**, type in a name for the new policy, then select a region where the policy will be located. It is recommended to use the same region where the vWAN and the secured hub are located.

5. For **Policy tier**, select **Standard** or **Premium** and click on **Review + create**:

Figure 7.21 – Creating an Azure Firewall policy

6. Click on **Create** after the validation has been successful. This process takes about 5 minutes to complete.

Adding a rule to the firewall policy

Now that you have a firewall policy, a rule will need to be added to it to make it effective:

1. From the newly created firewall policy, click on **Application rules** under **Settings**. Then, click on + **Add rule collection**.

2. On the **Add a rule collection** page, type in a name for your rule collection in the **Name** section.

3. For **Rule collection type**, select **Application**.

4. For **Priority**, type 100.

5. Select **Allow** for the **Rule collection action** option.

6. Select **DefaultApplicationRuleCollectionGroup** for the **Rule collection group** option.

7. Enter a name for the rules in the **Name** box.

8. Set the source type to **IP Address**.

9. For **Source**, type *. This will ensure that all segments in your AVS environment will be able to access the internet through the vWAN.

10. For **Protocol**, type http, https.

11. Select **FQDN** for the **Destination type** option.

12. Type * for **Destination**.

13. Click **Add**:

Figure 7.22 – Adding a rule to the Firewall Policy

Associating the Firewall policy with the hub

Now that you have added a rule to the firewall policy, you will need to associate it with the hub for the rule to be effective:

1. Go to the **Firewall Manager** page and click on **Azure Firewall Policies** under **Security**.

2. Select the check box for the policy you created earlier.

3. Select **Manage associations** and click on **Associate hubs**.

4. Select the check box next to your hub.

5. Click **Add**:

Figure 7.21 – Associating a firewall policy with a hub

Routing the AVS traffic to the vWAN hub

You need to ensure that the internet traffic gets routed through Azure Firewall. Follow these steps to do so:

1. From the **Firewall Manager** page, select **Virtual hubs**.

2. Select the hub you created earlier:

Figure 7.22 – Routing AVS traffic through the firewall

3.   Select **Security configuration** under **Settings** on the hub page.

4.   Under **Internet traffic**, select **Azure Firewall**. (Optionally, you can do the same for **Private traffic**.)

5.   Click **Save**.

6.   Click **OK** on the pop-up screen.

This process will take a few minutes to complete:

Figure 7.23 – Routing AVS traffic through the firewall

Important note

In a production environment, this action should be done during a change control window as it will route all traffic from your AVS environment to the internet with Azure Firewall as your next hop from the ExpressRoute gateway.

Notice that, under the **CONNECTIONS** section, **INTERNET TRAFFIC** is **Unsecured**. Once the process has been completed, you will notice that the connection status will be **Secured**:

Figure 7.24 – Internet traffic secured by Azure Firewall

Now that you have finished routing all the traffic through the vWAN hub, you can test the process from any virtual machine in your AVS environment.

You also have the option of viewing all the routes that are being routed through the vWAN hub:

1. In the hub, click on **Effective Routes** under **Routing**.

2. Under **Choose route tables**, select **Default**. The first 100 effective routes will be displayed.

You will now see all the segments from AVS plus the management address CIDR broken up into different networks:

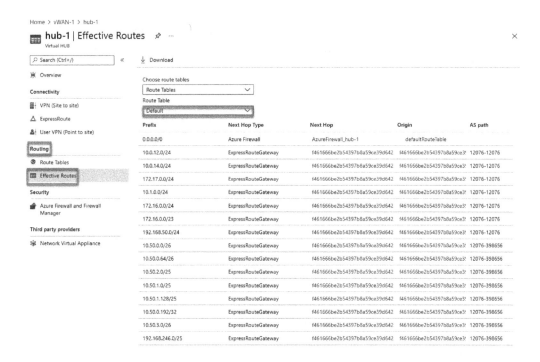

Figure 7.25 – Effective routes

Securing your internet traffic is very important as it provides you with a secure environment to operate in. You can take additional steps to ensure that traffic to and from your environment is safe. Some other steps include adding a web application firewall to filter your inbound application traffic and additional outbound web filtering so that only allowed websites are browsed from your environment. Each environment is different and configured differently but ensure that security is always the center of your architecture.

Summary

This chapter focused on securing the internet traffic from your AVS environment. While there are multiple options for doing so, the focus was on using a secured virtual WAN to secure the internet traffic.

Using a vWAN is one of the most common scenarios that customers use. Many customers already have a vWAN in their Azure environment and creating a new connection from the AVS environment is very straightforward.

You learned how to configure the different elements of a virtual WAN and how to make it secure using Azure Firewall.

In the next chapter, we will focus on implementing a **network virtual appliance (NVA)** for internet traffic inspection for your AVS environment. You will learn some critical reasons to ensure the traffic is inspected. You will also learn how to deploy and configure an Azure Route Server to integrate with the virtual network appliance.

8
Inspecting Traffic for AVS

When switching to the AVS, customers may want to preserve operational continuity with their existing third-party networking and security solutions (AVS). The communication mechanism has nothing to do with the NSX-T Service Insertion/Network Introspection certification process for vSphere or AVS, and third-party platforms may include products from Cisco, Juniper, Palo Alto Networks, and others.

In this chapter, we will take a closer look at the following topics:

- Internet consideration design options for AVS

- Implementing an NVA solution for traffic inspection

- Configuring the Route Server peering

At the time of writing, there are three main ways to provide inbound internet access to resources in your AVS environment and to create outbound access to the internet from AVS.

Those three options are the following:

- An existing internet service hosted in Azure

- **Source Network Address Translation (SNAT)** managed from AVS

- A public IP to the AVS NSX Edge

Internet consideration design options for AVS

There are many ways to create a default route in Azure and deliver it to your AVS environment. See the following choices:

- A third-party NVA in a native Azure virtual network coupled with an Azure Route Server

- A vWAN hub with an Azure firewall

- A default route from the customer on-premises environment transferred to AVS over Global Reach

- A third-party NVA in a vWAN hub-and-spoke Virtual Network configuration

Any of these patterns may be used to provide an outbound SNAT service, giving you the ability to choose which sources are permitted to leave the network, to see connection records, and, for certain services, to do further traffic inspection.

The same service can use an Azure Public IP and generate an incoming **Destination Network Address Translation (DNAT)** that points to targets in AVS.

It is also possible to create a system that uses many routes for internet traffic – one for incoming DNAT and another for outgoing SNAT, such as a third-party security NVA (such as a third-party load balancer NVA using SNAT pools for return traffic).

SNAT managed from AVS

An easy solution for outbound internet connectivity from an Azure VMware Solution private cloud is provided through a managed SNAT service. These are some of the features of this service:

- A SNAT gateway will provide all workload networks with instant outward connectivity to the internet when the radio button on the **Internet Connectivity** tab is selected

- All sources that access the SNAT service are permitted; there is no control over SNAT regulations

- No access to the connection logs

- Up to 128,000 simultaneous outbound connections are supported using two public IPs that are cycled

- The AVS Managed SNAT does not support inbound DNAT

Public IP to the AVS NSX Edge

This option sends an allotted Azure Public IP straight to the NSX Edge for use. It enables AVS to immediately apply and use public network addresses in NSX as necessary. The following connection types use these addresses:

- Inbound DNAT

- Outbound SNAT

- Load balancing using VMware AVI third-party NVAs

- Applications directly connected to a workload VM interface

With this choice, you may also establish a DMZ within AVS by configuring the public address on an NVA from a different vendor.

The included features are as follows:

- **Scale**: If an application needs more than the soft limit of 64 public IPs, 1,000s of additional public IPs may be requested and provided.

- **Flexibility**: You may use a Public IP anywhere in the NSX environment. On load balancers such as VMware's AVI or third-party NVAs, it may be utilized to offer SNAT or DNAT. It may also be applied to VMware segments, standalone VMs, or third-party network virtual security appliances.

- **Regionality**: Only the nearby SDDC's Public IP address may access the NSX Edge. They can both have a local exit if you have two or more AVS private clouds linked and have a Public IP set up. It is considerably simpler to direct traffic locally for a multi-private cloud in scattered regions with a local exit to internet intents than to attempt to manage default route propagation for a security or SNAT service hosted in Azure.

Some considerations for which option you choose to utilize

The following factors determine the choice you make:

- You're using an Azure native solution and configuring a default route from Azure to your AVS environment that inspects all internet traffic.

- You have two alternatives if you need to operate a third-party NVA to meet the current criteria for security inspection. You may use the default route technique to run your Public IP in Azure natively, or you can use the Public IP to NSX edge approach in AVS.

- The number of Public IPs that may be assigned to an NVA operating in native Azure or configured on an Azure Firewall is scaled. The Public IP to NSX edge option enables larger allocations (1,000s versus 100s).

- It might not be easy to match an AVS private cloud with an Azure security service when using numerous AVSs in different Azure regions that need to connect to the internet. This challenge stems from the way an Azure default route operates. For a localized exit to the internet from each private cloud in its local area, connect a Public IP to the NSX.

Implementing an NVA solution for traffic inspection

The NVA solution will need to have an Azure Route Server deployed in the same Azure virtual network. This is because an NVA cannot communicate directly with the Azure ExpressRoute gateway that is also needed for this solution to work. The following is a high-level architecture of the solution that will be detailed in this section:

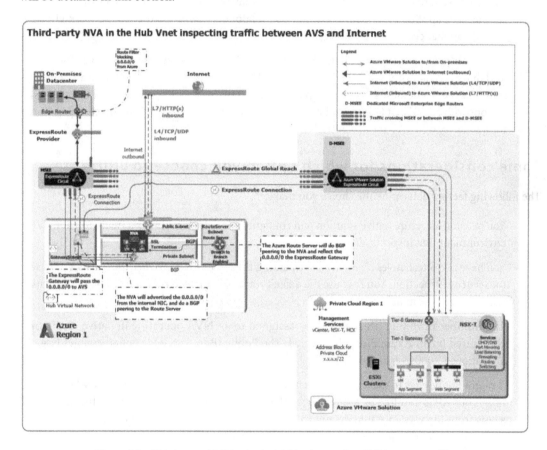

Figure 8.1 – Third-party NVA in the hub VNet inspecting AVS internet traffic

Figure 8.1 shows the Route Server, the ExpressRoute gateway, and the NVA in the same virtual network. However, the NVA can be in a different virtual network if you need it to be.

We will now walk through the steps for creating and configuring an Azure Route Server and a Quagga network virtual appliance.

Prerequisites

The prerequisites are as follows:

- An Azure subscription
- Minimum contributor access in the Azure subscription

Creating a virtual network

A virtual network is needed to deploy both the Azure Router Server and the Quagga NVA. A dedicated subnet is needed for each component:

1. Log into your Azure portal and click on + **Create a resource**. Then, in the search box, type in `virtual network` and press *Enter* on your keyboard:

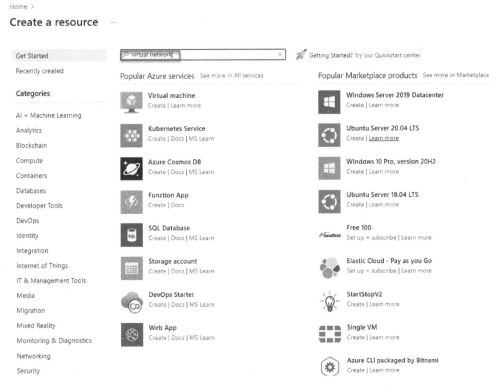

Figure 8.2 – Virtual network creation

2. Select **Create**.

3. Provide the following information on the **Basics** tab, then select **Next: IP Address >**:

Settings	Value
Subscription	Select the subscription that you will be using for this deployment
Resource Group	Select an existing resource group or create a new one
Name	Enter a name for the virtual network
Region	Select a region for which you will be deploying the virtual network – for example, Brazil South

Table 8.1 – Virtual network Basics tab settings

Refer to the following screenshot for this step:

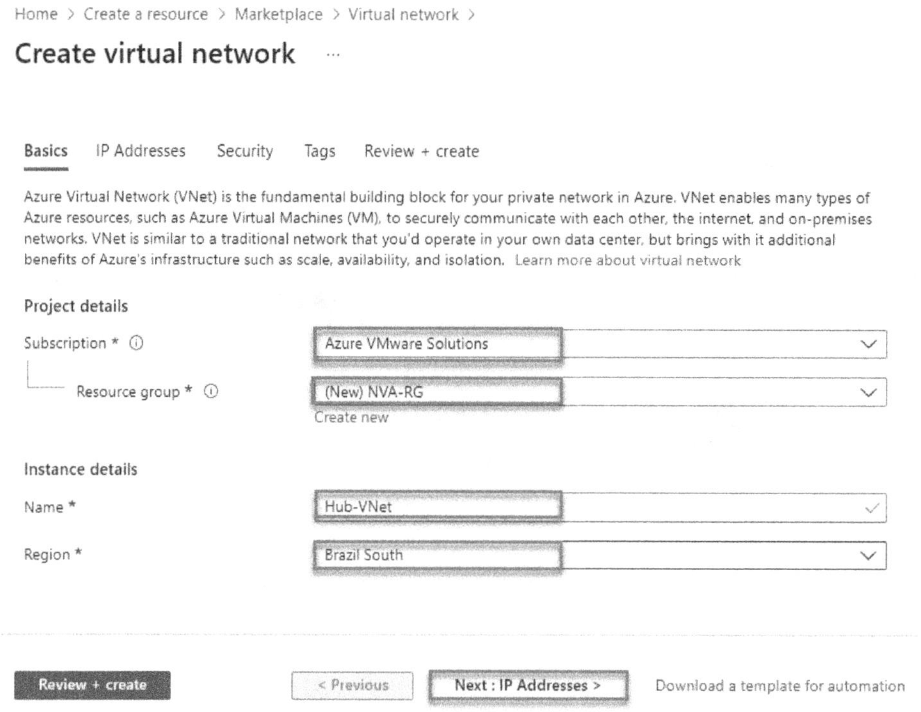

Figure 8.3 – Virtual network Basics tab information

4. On the **IP Address** tab, configure the virtual network address space as desired – for example, `172.16.0.0/16`. You will then need to create the individual subnets for each component. Remember that the Route Server and the ExpressRoute gateway need to be in their own subnets. You should also put the NVA in its own subnet.

5. Click on + **Add subnet** and enter a name for the subnet and fill out the subnet address range:

> **Important note**
> RouteServerSubnet must be named as it is here.

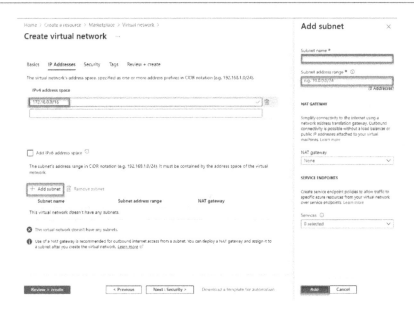

Figure 8.4 – Creating a new subnet

6. Click **Add**.

7. Repeat the steps shown in *Figure 8.4* for each new subnet:

Subnet Name	Subnet Address Range
RouteServerSubnet	172.16.1.0/25
GatewaySubnet	172.16.2.0/24
NVASubnet	172.16.3.0/24
Subnet1	172.16.4.0/24
Subnet2	172.16.5.0/24

Table 8.2 – Subnet details

8. After the new subnets have been created, click on **Review + create**. Optionally, you can choose to create a firewall, a bastion host, or enabled DDoS protection:

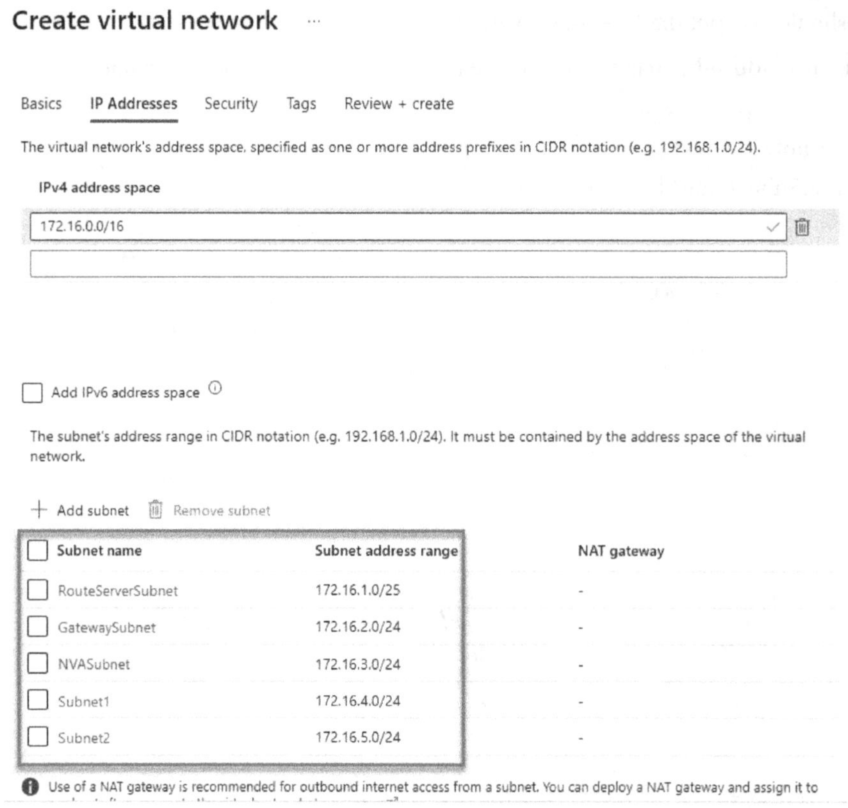

Create virtual network   ...

Basics   **IP Addresses**   Security   Tags   Review + create

The virtual network's address space, specified as one or more address prefixes in CIDR notation (e.g. 192.168.1.0/24).

IPv4 address space

172.16.0.0/16   ✓ 🗑

☐ Add IPv6 address space ⓘ

The subnet's address range in CIDR notation (e.g. 192.168.1.0/24). It must be contained by the address space of the virtual network.

+ Add subnet   🗑 Remove subnet

Subnet name	Subnet address range	NAT gateway
RouteServerSubnet	172.16.1.0/25	-
GatewaySubnet	172.16.2.0/24	-
NVASubnet	172.16.3.0/24	-
Subnet1	172.16.4.0/24	-
Subnet2	172.16.5.0/24	-

ⓘ Use of a NAT gateway is recommended for outbound internet access from a subnet. You can deploy a NAT gateway and assign it to

Figure 8.5 – New subnets created

9. Wait for the validation to pass and then click on **Create**. Wait for the deployment to be completed and then go to the next steps:

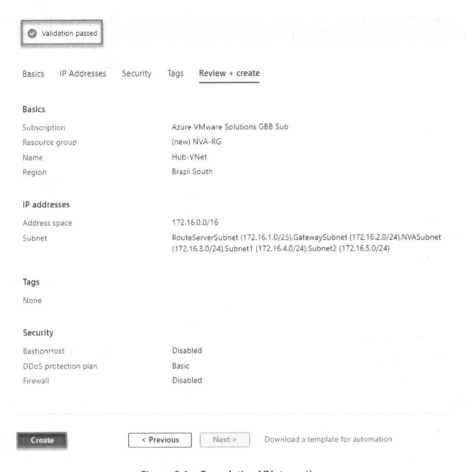

Create virtual network   ...

Validation passed

Basics   IP Addresses   Security   Tags   **Review + create**

Basics

Subscription   Azure VMware Solutions GBB Sub
Resource group   (new) NVA-RG
Name   Hub-VNet
Region   Brazil South

IP addresses

Address space   172.16.0.0/16
Subnet   RouteServerSubnet (172.16.1.0/25),GatewaySubnet (172.16.2.0/24),NVASubnet
                                 (172.16.3.0/24),Subnet1 (172.16.4.0/24),Subnet2 (172.16.5.0/24)

Tags

None

Security

BastionHost   Disabled
DDoS protection plan   Basic
Firewall   Disabled

Create   < Previous   Next >   Download a template for automation

Figure 8.6 – Completing VNet creation

Important note

Please note that, as a best practice, the Azure Route Server and the Quagga NVA should be created in the same virtual network as the existing ExpressRoute Gateway that is already connected to the AVS ExpressRoute circuit.

Deploying an Azure Route Server

You will now create the Azure Route Server. This Azure Route Server will be used to communicate with the NVA and the ExpressRoute gateway using a BGP peering connection:

1. Log into your Azure portal and click on + **Create a resource**. Then, in the search box, type in `route server` and press *Enter* on your keyboard.

2. Click on **Create**.

Settings	Value
Subscription	Select the same subscription that the virtual network was created earlier
Resource Group	Select an existing resource group or create a new one
Name	Enter a name for the Route Server – for example, myRouterServer
Region	Select the same region you created the virtual network in – for example, Brazil South
Virtual network	Select the virtual network you created earlier – that is, Hub-VNet
Subnet	Select the RouteServerSubnet (172.16.1.0/25) you created earlier
Public IP address	Select an existing Standard public IP or create a new one that will be used with the Route Server

On the **Create a Route Server** page, select or enter the following information:

Table 8.3 – Route Server deployment

3. Click on **Review + create**:

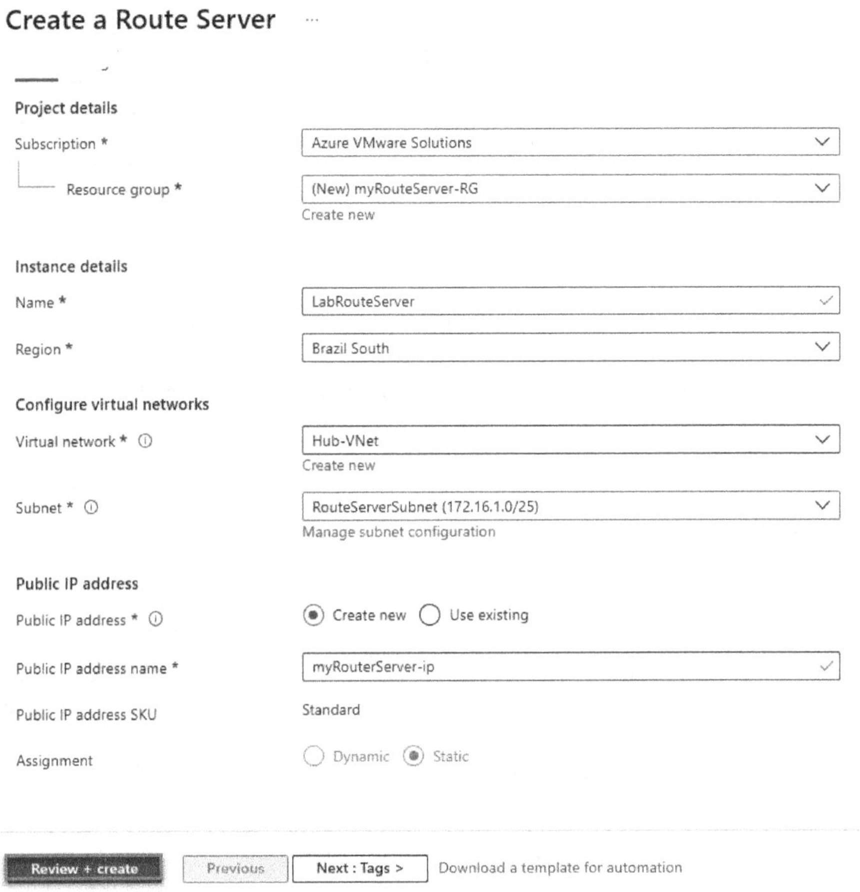

Create a Route Server   ...

Project details

Subscription * | Azure VMware Solutions   ✓ |

Resource group * | (New) myRouteServer-RG   ✓ |
                                       Create new

Instance details

Name * | LabRouteServer   ✓ |

Region * | Brazil South   ✓ |

Configure virtual networks

Virtual network * ⓘ | Hub-VNet   ✓ |
                                       Create new

Subnet * ⓘ | RouteServerSubnet (172.16.1.0/25)   ✓ |
                                       Manage subnet configuration

Public IP address

Public IP address * ⓘ   ◉ Create new ○ Use existing

Public IP address name * | myRouterServer-ip   ✓ |

Public IP address SKU   Standard

Assignment   ○ Dynamic ◉ Static

[Review + create]   [Previous]   [Next : Tags >]   Download a template for automation

Figure 8.7 – Route Server deployment

4. Wait for validation to pass and then click on **Create**.

Deploying a Quagga using an NVA

You have the option to use the NVA that you are most comfortable and familiar with. I will be deploying a Quagga NVA that will be configured on a Linux VM:

1. From your Azure portal, select **+ Create a resource**. Then, type `virtual machine` in the search box. Press *Enter*.

2. On the **Basics** tab, select or enter the following information as outlined. Make sure to use a strong password for the VM:

Settings	Value
Subscription	Select the same subscription that you deployed the virtual network with previously
Resource group	Select the existing resource group – that is, myRouteServer-RG
Virtual machine name	Enter the name Quagga
Region	Select the Brazil South region
Availability option	No infrastructure redundancy required
Security type	Standard
Image	Select Ubuntu 18.04 LTS - Gen 2
Azure Spot instance	Leave unchecked
Size	Select Standard_B2s - 2vcpus, 4GiB memory
Authentication type	Select Password
Username	Enter azureuser
Password	Enter and confirm the password of your choosing
Public inbound ports	Select Allow selected ports
Select inbound ports	Select SSH (22)

Table 8.4 – Network virtual appliance deployment

3. Click **Next: Disks >**:

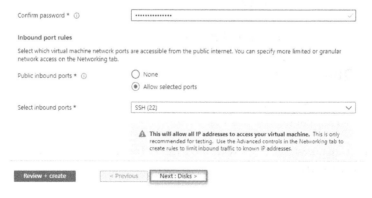

Figure 8.8 – Network virtual appliance deployment

4. On the **Disks** tab, accept the defaults and click on **Next: Networking >**:

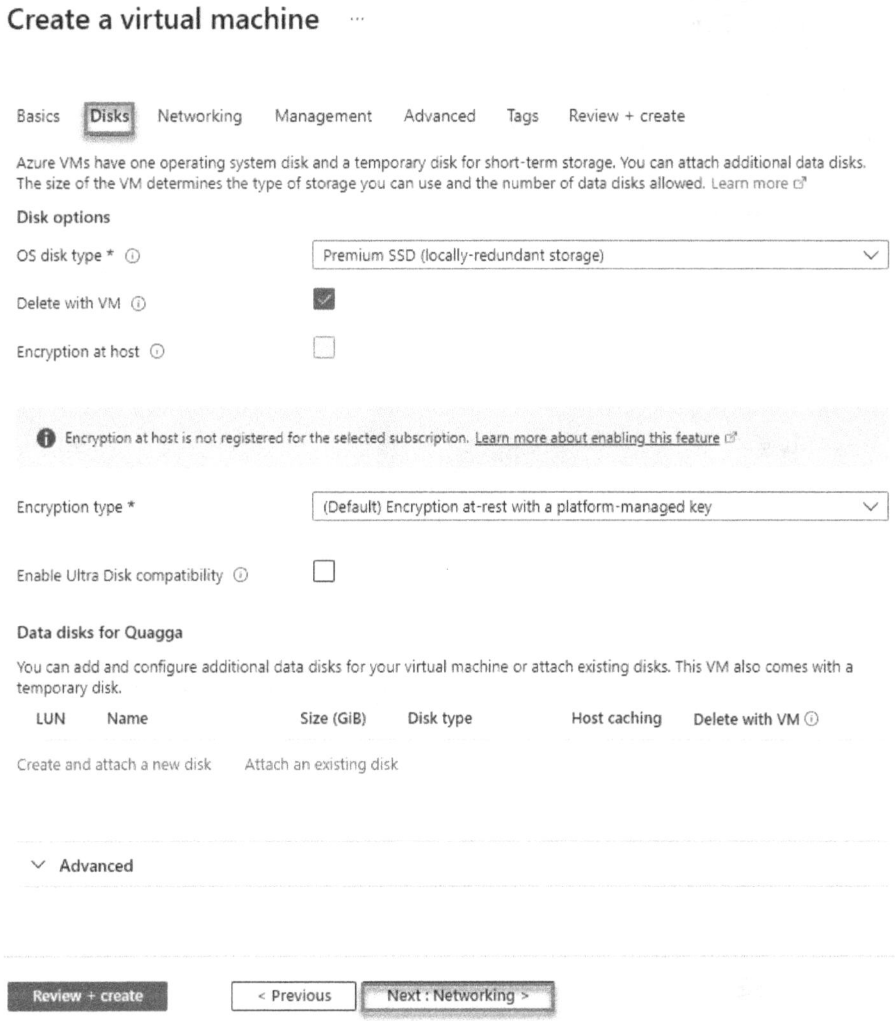

Create a virtual machine   ⋯

Basics   **Disks**   Networking   Management   Advanced   Tags   Review + create

Azure VMs have one operating system disk and a temporary disk for short-term storage. You can attach additional data disks. The size of the VM determines the type of storage you can use and the number of data disks allowed. Learn more ◱

Disk options

OS disk type * ⓘ | Premium SSD (locally-redundant storage)   ⌄ |

Delete with VM ⓘ   ☑

Encryption at host ⓘ   ☐

    ⓘ Encryption at host is not registered for the selected subscription. <u>Learn more about enabling this feature</u> ◱

Encryption type * | (Default) Encryption at-rest with a platform-managed key   ⌄ |

Enable Ultra Disk compatibility ⓘ ☐

Data disks for Quagga

You can add and configure additional data disks for your virtual machine or attach existing disks. This VM also comes with a temporary disk.

LUN	Name	Size (GiB)	Disk type	Host caching	Delete with VM ⓘ

Create and attach a new disk   Attach an existing disk

∨ Advanced

[Review + create]   [< Previous] [Next : Networking >]

Figure 8.9 – Network virtual appliance – Disks tab

5. On the **Networking** tab, select the virtual network that was created earlier (*Hub-VNet*) and then select the NVASubnet that was also created earlier. Accept the defaults for the other settings and then click on **Review + create**:

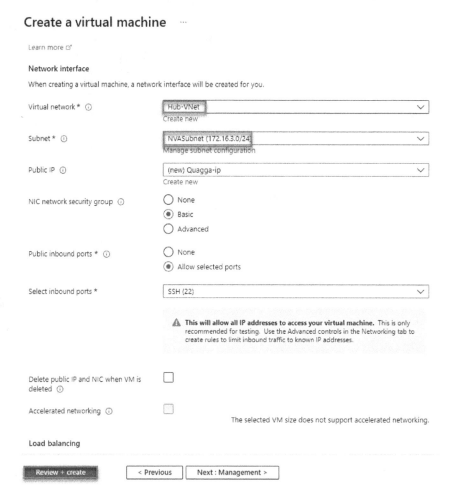

Figure 8.10 – Network virtual appliance network tab

6. Click **Create** after the validation has passed. The deployment will take about 10 minutes to complete.

7. After the VM has been deployed, go to the **Networking** settings of the VM and select the network interface:

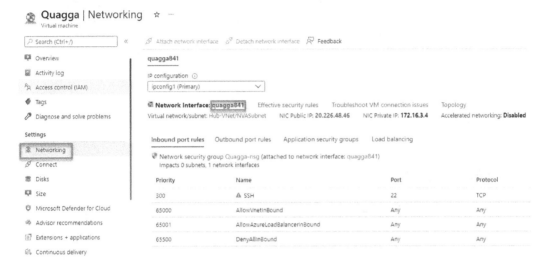

Figure 8.11 – Network interface of the NVA VM

8. Under **Settings**, select **IP configuration** and then select **ipconfig1**:

Figure 8.12 – IP configuration option

Take note of the private and public IP addresses for the VM. You will need those in the next steps.

9. Using PuTTY, connect to the VM using the public IP address and the credentials you used when you created the VM. (PuTTY is an SSH and telnet client, developed originally by Simon Tatham for the Windows platform. PuTTY is open source software that is available with source code and is developed and supported by a group of volunteers.)

10. Once you have logged in, enter sudo su to switch to superuser mode. You will need to copy the script located at https://raw.githubusercontent.com/Azure/azure-quickstart-templates/master/quickstarts/microsoft.network/route-server-quagga/scripts/quaggadeploy.sh and paste it into the PuTTY session. Please make sure that you modify the script based on your configuration. The script will configure the network virtual appliance along with other network settings.

Configuring the Route Server peering

The steps are as follows:

1. Go to the Azure Route Server you created earlier.

2. Under **Settings**, select **Peers** and select + **Add** to add a new peer:

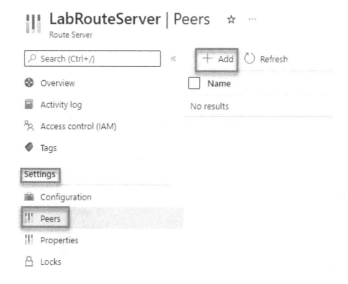

Figure 8.13 – Adding a Route Server peer

3. On the **Add Peer** page, enter the following information, and then click on **Add** to save the changes:

Setting	Value
Name	Use Quagga as the name.
ASN	Enter the ASN of the Quagga NVA. It is 65001.
Ipv4 Address	Enter the private IP address of the Quagga NVA virtual machine.

Table 8.5 – Adding a peer to the Route Server for the NVA VM

The screenshot for reference is as follows:

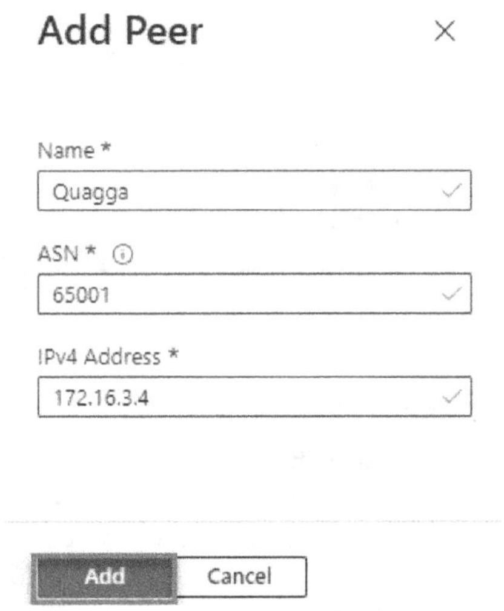

Add Peer   ✕

Name *

Quagga   ✓

ASN * ⓘ

65001   ✓

IPv4 Address *

172.16.3.4   ✓

Add   Cancel

Figure 8.14 – Route Server peering information

After the peering has been added to the Route Server, you should see it in the **Peers** section of the Router Server:

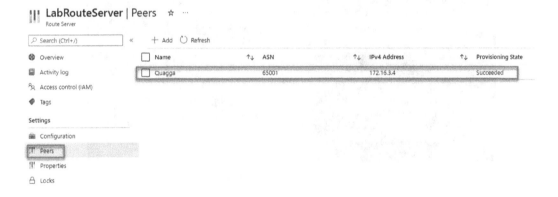

Figure 8.15 – Route Server new peering information

Now that the Route Server is in place and peering has been configured to the Quagga NVA, let's look at the learned routes.

Checking the learned routes on the Route Server and the Quagga NVA

You can check the routes that have been learned by the Route Server by using the following PowerShell command:

```
$routes = @{
     RouteServerName = 'LabRouteServer'
     ResourceGroupName = 'myRouteServer-RG'
     PeerName = 'Quagga'
}
Get-AzRouteServerPeerLearnedRoute @routes | ft
Please note that you may need to installthe
Az.Network PowerShell module in order to use the
Get-AzRouteServerPeerLearnedRoute command-let.
```

Adjust the command to suit your environment.

The output should look like this:

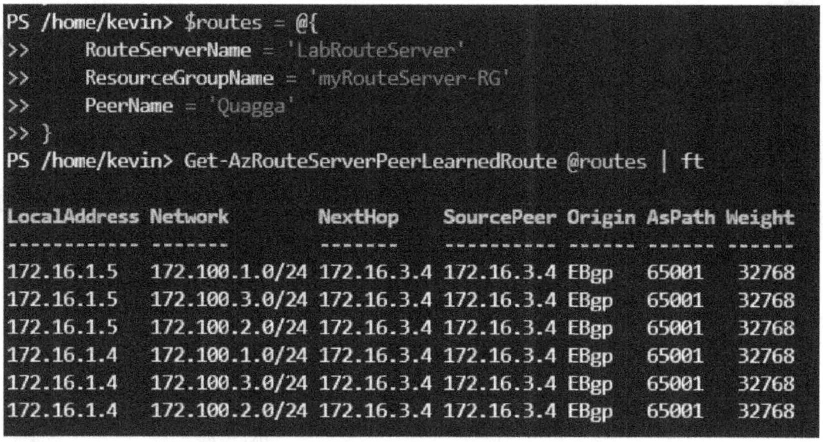

```
PS /home/kevin> $routes = @{
>>      RouteServerName = 'LabRouteServer'
>>      ResourceGroupName = 'myRouteServer-RG'
>>      PeerName = 'Quagga'
>> }
PS /home/kevin> Get-AzRouteServerPeerLearnedRoute @routes | ft

LocalAddress Network          NextHop    SourcePeer Origin AsPath Weight
------------ -------          -------    ---------- ------ ------ ------
172.16.1.5   172.100.1.0/24 172.16.3.4 172.16.3.4 EBgp   65001   32768
172.16.1.5   172.100.3.0/24 172.16.3.4 172.16.3.4 EBgp   65001   32768
172.16.1.5   172.100.2.0/24 172.16.3.4 172.16.3.4 EBgp   65001   32768
172.16.1.4   172.100.1.0/24 172.16.3.4 172.16.3.4 EBgp   65001   32768
172.16.1.4   172.100.3.0/24 172.16.3.4 172.16.3.4 EBgp   65001   32768
172.16.1.4   172.100.2.0/24 172.16.3.4 172.16.3.4 EBgp   65001   32768
```

Figure 8.16 – Route Server new peering information

`172.16.1.5` and `.4` are the IP addresses of the Route Server. The next hop IP address is that of the Quagga NVA.

To check the routes learned by the Quagga NVA, you will need to log into the NVA by using PuTTY. Use the public IP from the Quagga VM and log in using the username and password that was created earlier.

Once you have logged in, type vtysh and then enter show ip bgp. The output should look like this:

```
root@Quagga:/home/azuser# vtysh

Hello, this is Quagga (version 1.2.4).
Copyright 1996-2005 Kunihiro Ishiguro, et al.

Quagga# show ip bgp
BGP table version is 0, local router ID is 172.16.3.4
Status codes: s suppressed, d damped, h history, * valid, > best, = multipath,
              i internal, r RIB-failure, S Stale, R Removed
Origin codes: i - IGP, e - EGP, ? - incomplete

   Network          Next Hop         Metric LocPrf Weight Path
   10.0.12.0/24     172.16.1.5                        0 65515 12076 12076 i
                    172.16.1.4                        0 65515 12076 12076 i
   10.0.14.0/24     172.16.1.5                        0 65515 12076 12076 i
                    172.16.1.4                        0 65515 12076 12076 i
   10.1.0.0/24      172.16.1.5                        0 65515 12076 12076 i
                    172.16.1.4                        0 65515 12076 12076 i
   10.50.0.0/26     172.16.1.5                        0 65515 12076 398656 ?
                    172.16.1.4                        0 65515 12076 398656 ?
   10.50.0.64/26    172.16.1.5                        0 65515 12076 398656 ?
                    172.16.1.4                        0 65515 12076 398656 ?
   10.50.0.192/32   172.16.1.5                        0 65515 12076 398656 ?
                    172.16.1.4                        0 65515 12076 398656 ?
   10.50.1.0/25     172.16.1.5                        0 65515 12076 398656 ?
                    172.16.1.4                        0 65515 12076 398656 ?
   10.50.1.128/25   172.16.1.5                        0 65515 12076 398656 ?
                    172.16.1.4                        0 65515 12076 398656 ?
   10.50.2.0/25     172.16.1.5                        0 65515 12076 398656 ?
                    172.16.1.4                        0 65515 12076 398656 ?
   10.50.3.0/26     172.16.1.5                        0 65515 12076 398656 ?
                    172.16.1.4                        0 65515 12076 398656 ?
   172.16.0.0       172.16.1.5                        0 65515 i
                    172.16.1.4                        0 65515 i
*> 172.100.1.0/24   0.0.0.0               0       32768 i
*> 172.100.2.0/24   0.0.0.0               0       32768 i
*> 172.100.3.0/24   0.0.0.0               0       32768 i
   172.250.16.0/23  172.16.1.5                        0 65515 12076 12076 i
                    172.16.1.4                        0 65515 12076 12076 i
   192.168.0.0/22   172.16.1.5                        0 65515 12076 12076 i
                    172.16.1.4                        0 65515 12076 12076 i
   192.168.4.0      172.16.1.5                        0 65515 12076 12076 i
                    172.16.1.4                        0 65515 12076 12076 i
   192.168.50.0     172.16.1.5                        0 65515 12076 12076 i
                    172.16.1.4                        0 65515 12076 12076 i
   192.168.246.0/25 172.16.1.5                        0 65515 12076 398656 ?
                    172.16.1.4                        0 65515 12076 398656 ?
   192.168.247.0/25 172.16.1.5                        0 65515 12076 398656 ?
                    172.16.1.4                        0 65515 12076 398656 ?
   192.168.248.0/25 172.16.1.5                        0 65515 12076 398656 ?
                    172.16.1.4                        0 65515 12076 398656 ?
```

Figure 8.17 – Quagga NVA learned routes

The IP address of 172.16.0.0 is the network where we have all the devices configured. The next hop IP address is that of the Route Server.

The IP address of 10.50.0.0 is that of the AVS management network. Now, we can see that traffic from the AVS environment is routed to the Quagga NVA from the Azure Route Server.

With the preceding configuration, the default route to the internet from the AVS environment will be through the Quagga NVA. Since every customer is unique, you must configure the security requirements on the NVA as needed. As I mentioned earlier in this chapter, there are many solutions that customers can use to do traffic inspection of both their internet ingress and egress traffic.

Summary

In this chapter, we looked at how you can use an NVA (located in your Azure native environment) to inspect all traffic to and from the internet to AVS. We used the Quagga NVA in this chapter for traffic inspection, but you can use any third-party NVA.

You learned how to do the following:

- Integrate AVS with your Azure environment while utilizing your existing security solutions. This seamless approach works without introducing new solutions or technologies that are intended to protect your AVS environment.

- Deploy and configure an Azure Route Server, which is used to redirect traffic between your existing ExpressRoute Gateway and your virtual network appliance.

- Deploy and configure a Quagga NVA to integrate with your ExpressRoute Gateway and the Azure Route Server.

- View the learned routes on the Quagga NVA and the Azure Route Server.

In the next chapter, we will learn how to add additional storage to the AVS data store without adding additional nodes. We will look at the need to expand the data store and the different options to do so.

9
Storage Concepts in AVS

You can use Azure Storage resources to extend the storage capabilities of your private clouds. AVS provides a native, cluster-wide storage solution using VMware vSAN. The local storage from each host in a cluster is utilized in a vSAN data store, and data-at-rest encryption is enabled by default. To date, the vSAN storage is the fastest storage available in Azure today.

An all-flash VMware vSAN software-defined storage system is used as the storage in the AVS hyper-converged vSphere cluster. When utilizing AVS, the sophisticated software-defined storage solution vSAN offers provides several fantastic benefits.

Throughout this chapter, we will look at the following areas regarding AVS storage:

- vSAN clusters
- Fault tolerance and storage policies
- Configuring a storage policy
- Azure NetApp Files

Each cluster host's local storage is claimed as a component of a vSAN data store. With a raw, per host, SSD-based capacity of 15.4 TB, all disk groups employ an NVMe cache layer of 1.6 TB. The per-host capacity multiplied by the number of hosts determines the size of a cluster's raw capacity tier. For instance, the vSAN capacity tier offers 61.6 TB of raw capacity for a cluster of four hosts.

When a customer is doing a sizing exercise to determine the number of AVS nodes needed to accommodate their workload, it is crucial to ensure the number of hosts will accommodate the required storage.

Every data store is built as part of the AVS deployment and is instantly usable. In the cluster-wide vSAN data store, local storage in cluster hosts is used. These vSAN rights are available for usage by the `cloudadmin` user and all other users allocated to the CloudAdmin role to administer data stores:

- `Datastore.DeleteFile`
- `Datastore.FileManagement`

- `Datastore.UpdateVirtualMachineMetadata`
- `Datastore.AllocateSpace`
- `Datastore.Browse`
- `Datastore.Config`

> **Important note**
>
> You can't change the name of data stores or clusters once they have been deployed.

Understanding the storage policies and fault tolerance available in AVS is imperative. In the next section, we will look at the different options available in the solution.

Fault tolerance and storage policies

RAID-1 (Mirroring) FTT-1 is the default storage policy, while thin provisioning is the object space reservation setting. The cluster will continue using this default storage policy unless it is changed, or a new policy is applied.

> **Important note**
>
> You could see a VM storage policy named vSAN Default Storage Policy with Object Space Reservation set to thick provisioning when you log in to the vSphere Client. Please note that the cluster does not use this as its default storage policy. This rule is still in effect for historical reasons, although thin provisioning will soon replace it.
>
> The Microsoft vSAN Management Storage Policy is used by all of the **software-defined data center** (**SDDC**) management VMs (vCenter, NSX manager, NSX controller, NSX edges, and others), with Object Space Reservation set to thick provisioning.

Configuring a storage policy

VMware vSAN storage policies determine your virtual machines' storage needs. Because they control how storage is assigned to the VM, these rules provide the necessary level of service for your virtual machines. At least one VM storage policy is given to each VM deployed to a vSAN data store.

When a VM is first deployed or whenever you do additional VM activities, such as cloning or migrating, you may set a VM storage policy. `cloudadmin` users or other roles cannot change the default storage policy for a VM with equal privileges after deployment. Changes to the VM storage policy per disk are allowed, nevertheless.

The Run command enables authorized users to modify the pre-existing or default VM storage policy to a different policy that is accessible for a VM after deployment. The disk-level VM storage policy remains unchanged. You may always modify the VM storage policy at the disk level to suit your needs.

We will now walk through the process of how to do the following:

- List all storage policies
- Set a storage policy for a VM
- Specify the default storage policy for an AVS cluster

Let's get started.

Prerequisites

The minimum level of hosts must be met to provide the respective number of disk **failures to tolerate (FTT)**.

Listing all storage policies

You will need to use the Run command's `Get-StoragePolicy` cmdlet to list the vSAN-based storage policies that are available to be set on any VM. Follow these steps:

1. Go to your AVS portal.
2. Under **Operations**, select **Run command | Packages | Get-StoragePolicies**:

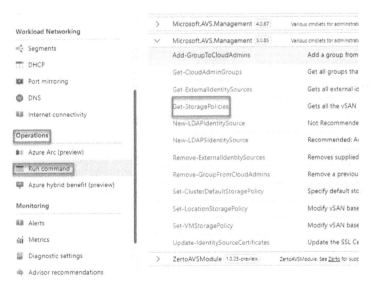

Figure 9.1 – The Get-StoragePolicies cmdlet

3. Keep the default values or specify new ones and then select **Run**:

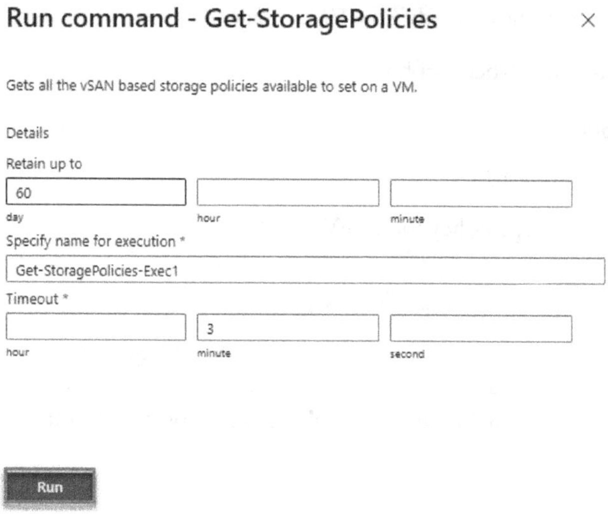

Run command - Get-StoragePolicies   ✕

Gets all the vSAN based storage policies available to set on a VM.

Details

Retain up to

60		
day	hour	minute

Specify name for execution *

Get-StoragePolicies-Exec1

Timeout *

	3	
hour	minute	second

Run

Figure 9.2 – Run command input fields

The following table lists the Get-StoragePolicies cmdlet's field inputs:

Field	Value
Retain up to	The retention period for the cmdlet's output. 60 is the default.
Specify name for the execution	This is an alphanumeric name; for example, Get-StoragePolicies-Exec2.
Timeout	This is the period after which the cmdlet exits if it is taking too long to finish up.

Table 9.1 – The Get-StoragePolicies cmdlet's field input

4. To view the output of the cmdlet, click on **Run execution status**.

5. Click on **Get-StoragePolicies-Exec1**:

Figure 9.3 – Getting the status of the Run command

6. Click on **Output** to view the available storage policies:

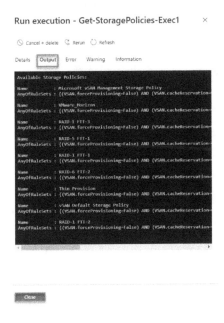

Figure 9.4 – Available storage policies

7.   Click **Close** whenever you are done viewing the storage policies.

Setting a storage policy for a VM

For this task, you will need to run the `Set-VMStoragePolicy` cmdlet to modify the vSAN-based storage policy on the default cluster, individual VM, or group of VMs sharing a similar VM name. For example, if you have two VMs named `XYZVM1` and `XYZVM2`, entering `XYZVM` in the `VMName` parameter would change the storage policy for both VMs:

> **Important note**
>
> The vSphere Client cannot be used to change the default storage policy or any existing policy for a VM. You will need to use the `run` cmdlet.

1.   To set the storage policy for a VM, select **Run command** | **Packages** | **Set-VMStoragePolicy**:

Figure 9.5 – Set-VMStoragePolicy

2.   Fill in the required values and then select **Run**:

Field	Value
VMName	The name of the VM that you will be making the change on.
StoragePolicyname	The name of the storage policy you will be setting; for example, RAID-1 FTT-1.
Retain up to	The retention period for the cmdlet output. 60 is the default value.
Specify name for the execution	The alphanumeric name for this field.
Timeout	This is the period after the cmdlet exists if it is taking too long to be completed.

Table 9.3 – The Set-VMStoragePolicy field

3.   After you enter the required information, click **Run**:

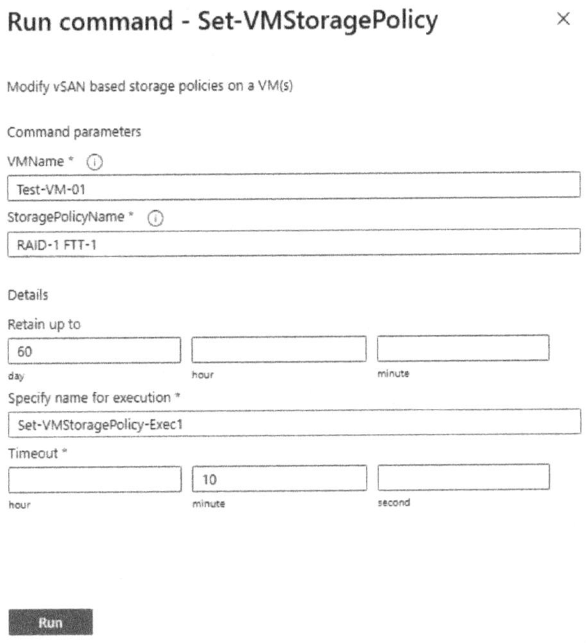

Run command - Set-VMStoragePolicy   ✕

Modify vSAN based storage policies on a VM(s)

Command parameters

VMName *   ⓘ

| Test-VM-01 |

StoragePolicyName *   ⓘ

| RAID-1 FTT-1 |

Details

Retain up to

60		
day	hour	minute

Specify name for execution *

| Set-VMStoragePolicy-Exec1 |

Timeout *

	10	
hour	minute	second

Run

Figure 9.6 – Set-VMStoragePolicy fields

4. Once the cmdlet has successfully run, click on **Run execution status**. Then, select **Set-VMStoragePolicy-Exex1**:

Figure 9.7 – Checking the output of Set-VMStoragePolicy

5. Click on **Details** to view the status of the cmdlet:

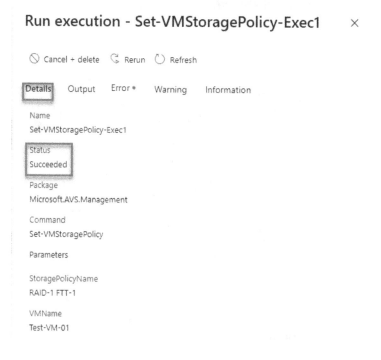

Figure 9.8 – Successful Set-VMStoragePolicy cmdlet

6. Click on the **Output** tab to see the new storage policy for the VM:

Figure 9.9 – Successful Set-VMStoragePolicy cmdlet

7. Click **Close**.

Specifying the default storage policy for an AVS cluster

We will now go through the steps of setting a storage policy for a specific cluster by using the `Set-ClusterDefaultStoragePolicy` cmdlet. To do so, follow these instructions:

1. Select **Run command** from the **Operations** section on your AVS portal.

2. Click on **Packages | Set-ClusterDefaultStoragePolicy**:

Figure 9.10 – Set-ClusterDefaultStoragePolicy

3. Provide the required values, as listed in the following table, or keep the default values:

Field	Value
ClusterName	The name of the cluster.
StoragePolicyName	The name of the storage policy you will be setting.
Retain up to	The retention period for the cmdlet to show output. 60 is the default value.
Specify name for execution	This is an alphanumeric value for the name of the execution.
Timeout	The length of time before the cmdlet exits if it is taking too long to execute.

Table 9.4 – Set-ClusterDefaultStoragePolicy field values

4. Click **Run**:

Figure 9.11 – Set-ClusterDefaultStoragePolicy value fields

5. Wait for the command to complete. Then, click on the **Run execution status** tab.

6. Click on the **Set-ClusterDefaultStoragePolicy-Exec1** execution name:

Figure 9.12 – Set-ClusterDefaultStoragePolicy-Exec1 status

7. Click on the **Output** tab to see the status of the run command:

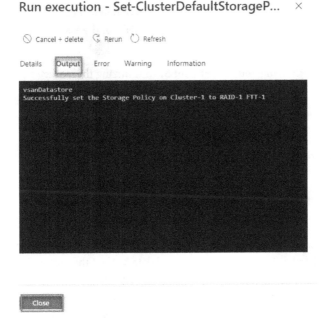

Figure 9.13 – Set-ClusterDefaultStoragePolicy output status

As you have seen, there are techniques to optimize the use of the disk, just as with any storage system. The storage policies of each VM may be customized using VMware vSAN. The level of granularity extends well beyond the VM; each disk connected to the VM might have a unique storage policy.

Any disk generated in the private cloud of AVS will have a default storage policy of 1 failure – RAID-1 (Mirroring). Therefore, even if one host fails, no data will be lost. In this setup, twice as much raw disk is required to sustain the used disk. This is the sole viable policy when there are just three nodes in a vSAN cluster. An AVS private cloud often starts with the minimum three (3)-node setup before starting to migrate virtual machines from on-premises to AVS or as workloads expand naturally. The three-node cluster will eventually need to be extended to a fourth, fifth, or sixth node, and so on, as more and more VMs start to fill the cluster. The beauty of cloud-scale computing, and specifically, running VMware on the cloud, is the cluster's capacity to expand and contract as needed. On-premises VMware clusters are often overprovisioned, which creates a unique set of difficulties. With AVS, there is no need to do that.

Additional storage policies become available as the cluster expands.

Please note that as the cluster expands, the vSAN storage will soon be maxed out if the initial storage policy applied to the virtual machine disks stays the same. Compared to the RAID 1 setup, RAID 5/6 policies provide much more efficient use of storage.

Select the storage policy that works best for the VM disks when the AVS private cloud expands beyond three nodes. Reconfigure the storage rules on the disks of the VMs installed when the cluster has three nodes.

By doing this, you are maximizing your investment by minimizing the storage use of the AVS cluster.

In the next section, we will look at how the data on the AVS vSAN is encrypted. Microsoft uses encryption of data at rest for the data that resides on AVS vSAN.

Encryption of data at rest

Data-at-rest encryption is used by default in the vSAN data stores by utilizing keys kept in Azure Key Vault. The encryption program is KMS-based and works with vCenter Server key management functions. All data on SSDs is invalidated instantly upon a host's removal from a cluster.

Encryption at rest prevents an attacker from accessing unencrypted data by encrypting it on a disk. If an attacker obtains a hard disk containing encrypted data but not the encryption keys, the attacker must decrypt the data to access the information. This attack is significantly more complicated and resource-intensive than obtaining unencrypted hard disk data. Encryption at rest is therefore strongly recommended and a requirement of high importance for many companies.

Azure NetApp Files

As a persistent storage choice, **Network File System** (**NFS**) data stores are supported by AVS. With Azure NetApp Files volumes, you can build NFS data stores and join them to any cluster you want. **Virtual machines** (**VMs**) may also be built for the best cost and performance.

With minimal code modifications, databases and high-performance computing applications may be moved and operated in Microsoft Azure using Azure NetApp Files, an Azure service. Running within VMs, the guest operating system has access to Azure NetApp Files volumes. As a file share for workloads running on AVS, the Azure NetApp Files volume will be configured, tested, and verified in this section utilizing the NFS protocol.

The same Azure region is used to produce AVS and Azure NetApp Files. Numerous Azure regions provide Azure NetApp Files, which allow replication across different regions.

Azure NetApp Files supports the following services:

- **Azure VMware Solution**: Azure NetApp Files can be used to create NFS data stores and have them mounted on your AVS clusters. VMs built in the AVS environment may mount Azure NetApp Files shares.

- **Share Protocol**: The **Server Message Block** (**SMB**) and NFS protocols are supported by Azure NetApp Files. Because of this functionality, the volumes may be mapped to Windows clients and mounted on Linux clients.

- **Active Directory connections**: Active Directory Domain Services and Azure Active Directory Services are also supported by Azure NetApp Files.

Prerequisites

The following are required to create a NetApp File volume:

- An Azure subscription with the Azure NetApp File resource provider registered
- A subnet delegated for Azure NetApp Files

Creating a NetApp Files volume for AVS

In this section, you will learn how to create and mount a NetApp Files volume for an AVS virtual machine.

We will walk through the following processes:

- Creating a NetApp Account
- Setting up a capacity pool
- Delegating a subnet to Azure NetApp Files

- Creating an NFS volume for Azure NetApp Files

Let's get started.

Creating a NetApp account

When you create a NetApp account, you can set up a capacity pool and then create a volume. The **Azure NetApp Files** blade is used to create a new NetApp account:

1. Sign in to the Azure portal.

2. In the search bar, type in `Azure NetApp Files` and click on it when it shows up.

3. Click **+ Create** to create a new NetApp account:

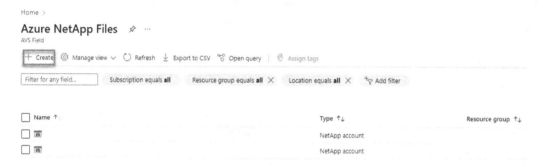

Figure 9.14 – Creating an Azure NetApp Files account

4. Fill in the required information and click **Create**:

   - **Name**: Specify a unique name for the subscription

   - **Subscription**: Select a subscription from your existing subscriptions

   - **Resource group**: Use an existing resource group or create a new one

- **Location**: Select the region where you want the NetApp account and its child resources to be located:

Figure 9.15 – New Azure NetApp Files account creation

After the process has been completed, the new NetApp Files account will be displayed in the **Azure NetApp Files** blade:

Figure 9.16 – New Azure NetApp Files account

Creating a capacity pool for Azure NetApp Files

The steps are as follows:

1. Go to the NetApp account you created and click on **Capacity pools** under **Storage service**:

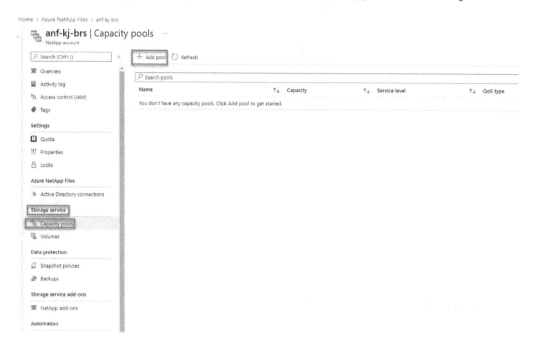

Figure 9.17 – Creating a capacity pool in the new NetApp Files account

2. Click on + **Add pool**.

3. Provide the following information for the new capacity pool:

   - **Name**: Specify a unique name for the capacity pool. Each capacity pool name must be unique for each NetApp account.

   - **Service level**: This field shows the target performance for the capacity pool. Specify the service level for the capacity pool: the options are **Ultra**, **Premium**, or **Standard**.

   - **Size**: Specify the size in TiB of the capacity pool that you are purchasing. The minimum capacity pool size is 4 TiB. You can change the size of a capacity pool in 1-TiB increments whenever the need is there.

   - **QoS**: Specify whether the capacity pool should use the **Manual** or **Auto** QoS type.

4. Click **Create**:

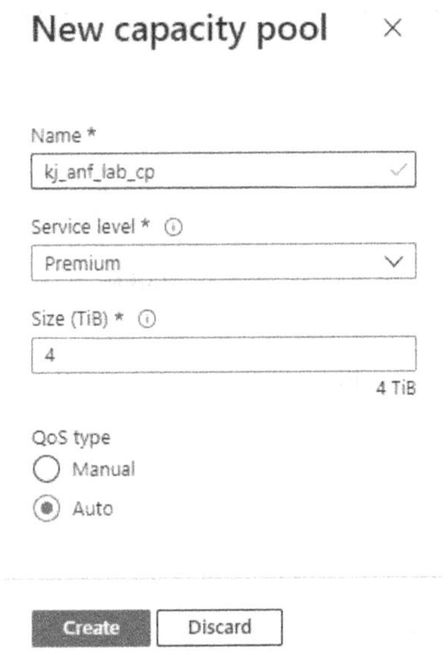

Figure 9.18 – New capacity pool deployment

Once completed, the new capacity pool will be listed under the **Capacity pools** tab. You will notice that the service level and the size of the storage pool are also shown:

Figure 9.19 – New capacity pool

Delegating a subnet to Azure NetApp Files

A subnet needs to be delegated to the Azure NetApp Files volume. When you are creating a volume, you will be prompted to provide a delegated subnet.

Some considerations are as follows:

- The size of the subnet depends on how many storage volumes and storage endpoints you plan on using.

- Only one subnet in a VNet is allowed to be delegated to Azure NetApp Files.

- You will not be able to designate a network security group or service endpoints in the delegated subnet. Doing so will break the subnet delegation.

- You can create a new VNet with a new subnet for delegation or you can use an existing VNet and a subnet. You can also create a new subnet for delegation.

- You can delegate a subnet before creating the volume or you can create and delegate the subnet when you are creating the volume.

- Access to a volume from a globally peered virtual network is not currently supported.

- To establish routing or access control to the Azure NetApp Files-delegated subnet, you can apply UDRs and NSGs to other subnets, even within the same VNet as the subnet delegated to Azure NetApp Files.

- For Azure NetApp Files support of **user-defined routes (UDRs)** and **network security groups (NSGs)**, see *Constraints* in *Guidelines for Azure NetApp Files network planning* (https://learn.microsoft.com/en-us/azure/azure-netapp-files/azure-netapp-files-network-topologies#constraints).

We will create and delegate the subnet during the volume creation process in the next section.

Creating an NFS volume for Azure NetApp Files

Azure NetApp Files supports volumes that are using NFS (NFSv3, NFSv4) or dual protocol (NFSv3 and SMB, or NFSv4.1 and SMB). SMB3 is also supported.

In this section, we will show you how to create an NFS volume while creating a new subnet for delegation in an existing VNet.

The prerequisites are as follows:

- A capacity pool needs to be created

- A subnet must be delegated to Azure NetApp Files (this will be done in the volume creation process)

Follow these steps to create an NFS volume:

1. Click on **Volumes** in the **Storage service** section of the NetApp Files account.

2. Click on + **Add volume**:

Figure 9.20 – Creating a new NFS volume

3. In the **Create a volume** section, provide information for the following fields under the **Basic** tab:

    •   **Volume name**: Specify the name of the volume that you are creating. Each capacity pool's volume names must be distinct. A minimum of three characters must be used. Letters are required to start the name. Only characters other than underscores (_) and hyphens (-) are permitted in it. You cannot use `default` or `bin` as the name of the volume.

    •   **Capacity pool**: Specify which capacity pool the volume will be created in.

    •   **Quota**: Specify the amount of logical storage that should be allocated to the volume. The amount of free space in the selected capacity pool that may be utilized to create a new volume is shown in the **Available quota** field. The new volume's size must not go beyond the allotted limit.

    •   **Throughput (MiB/S)**: Specify the desired throughput for the volume if it is being created in a manual QoS capacity pool. For this exercise, the throughput field is grayed out because the volume is being created in a capacity pool that was created with the QoS set to auto.

    •   **Virtual network**: Select the Azure virtual network from which you want to access the volume. The VNet you select must have a subnet delegated to Azure NetApp Files. You can also choose the subnet for delegation while creating the volume, as will be done for this exercise.

- **Subnet**: Select a delegated subnet that you want to use for the volume. If you have not yet delegated a subnet for Azure NetApp Files, you can click **Create new** on the **Create a volume** page. On the **Create Subnet** page, specify the subnet information. **Microsoft.NetApp/ volumes** will be automatically selected as the default subnet delegation. Only one subnet can be delegated to Azure NetApp Files in each VNet:

Figure 9.21 – Creating a new delegated subnet for ANF

The following screenshot is for the entirety of *step 3*:

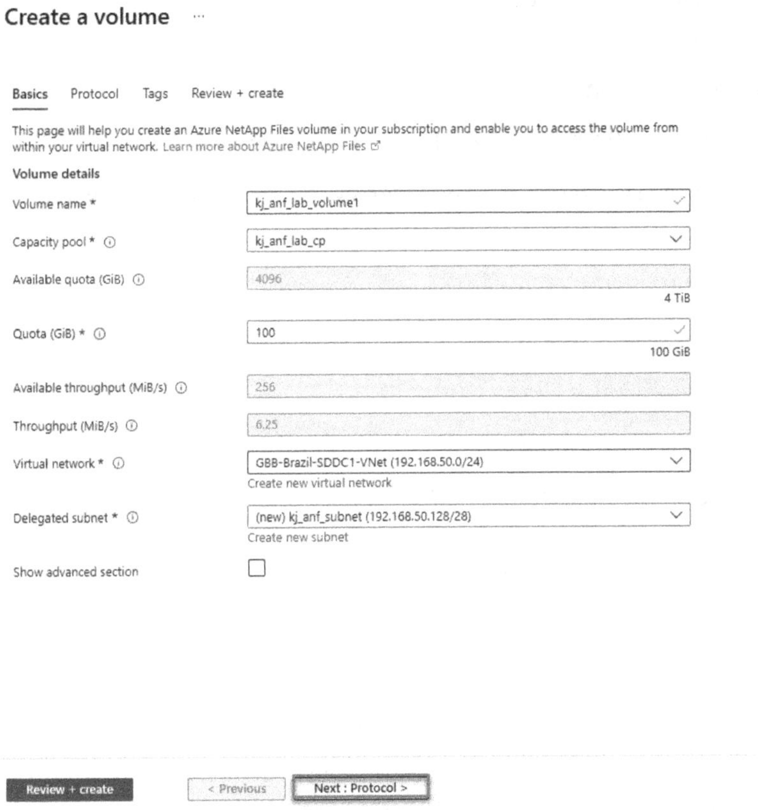

Create a volume   ...

Basics   Protocol   Tags   Review + create

This page will help you create an Azure NetApp Files volume in your subscription and enable you to access the volume from within your virtual network. Learn more about Azure NetApp Files ⧉

Volume details

Volume name *	kj_anf_lab_volume1
Capacity pool * ⓘ	kj_anf_lab_cp
Available quota (GiB) ⓘ	4096
	4 TiB
Quota (GiB) * ⓘ	100
	100 GiB
Available throughput (MiB/s) ⓘ	256
Throughput (MiB/s) ⓘ	6.25
Virtual network * ⓘ	GBB-Brazil-SDDC1-VNet (192.168.50.0/24)
	Create new virtual network
Delegated subnet * ⓘ	(new) kj_anf_subnet (192.168.50.128/28)
	Create new subnet
Show advanced section	☐

Review + create   < Previous   Next : Protocol >

Figure 9.22 – New volume information

4. Click on **Next: Protocol >**.

5. Select **NFS** as the protocol for this volume.

6. Specify a unique name for the file path. This will be used to create mount targets. There are some restrictions regarding the path names:

   - It must start with an alphabetical character

   - The name must be unique within each subnet in the region

   - It can contain only numbers, letters, or dashes (-)

   - It cannot exceed 80 characters

7.   Accept the **NFSv3** protocol for this volume.

8.   Leave the default disabled option for **LDAP**.

9.   Check the box for **Azure VMware Solution Datastore**.

10. Click on **Review + create**.

11. Click on **Create** once the validation has passed. This process takes about 10 minutes to complete.

The newly created volume will be listed in the **Volumes** section, as seen in the following screenshot:

Figure 9.23 – New ANF volume

We now have an Azure NetApp Files volume that can be attached to AVS hosts. At the time of writing this book, this solution is currently in Public Preview.

In the next section, we will walk you through the process of attaching an ANF volume to an AVS cluster.

Attaching an Azure NetApp Files volume to an AVS cluster

You can increase your AVS storage without scaling the clusters by utilizing NFS data stores supported by Azure NetApp Files. This is a very good option for customers who need additional storage but do not want to add additional AVS nodes to their cluster.

We will walk through the process of attaching the volume that was created to the AVS cluster.

The prerequisites are as follows:

- Deploy an AVS private cloud and a dedicated virtual network connected via ExpressRoute gateway.

- The virtual network gateway should be configured with the Ultra performance SKU and have FastPath enabled. An Azure NetApp Files volume must be created with the NFSv3 protocol. This should be configured in the same VNet that is connected to the AVS environment.

- Make sure that, when the volume was created, the **Azure VMware Solution Datastore** option was selected.

- Verify that you're registered for both the **CloudSanExperience** and **AnfDatstoreExperience** features by going to **Subscriptions | Select Subscription | Resource providers | Search for Microsoft.AVS | Register | Settings | Preview features**.

- Verify connectivity from the private cloud to the Azure NetApp Files volume by pinging the attached target IP.

Supported regions

Azure NetApp Files is currently supported in the following regions:

- **North America**: Canada Central, Canada East, Central US, East US, East US 2, North Central US, South Central US, West US, and West US 2

- **Brazil**: Brazil South

- **Europe**: France Central, Germany West Central, North Europe, Sweden Central, Sweden North, Switzerland West, UK South, UK West, and West Europe

- **Australia**: Australia East and Australia Southeast

- **Asia**: East Asia, Japan East, Japan West, and Southeast Asia

Performance best practices

NFS data stores on Azure NetApp Files volumes should adhere to some crucial best practices for maximum performance:

- Create Azure NetApp Files volumes with Standard network features to enable optimum connectivity from AVS through ExpressRoute FastPath.

- Select the appropriate service level for the Azure NetApp Files capacity pool based on your performance requirements. The Ultra tier is suggested for the best performance.

- Choose the **UltraPerformance** gateway and enable ExpressRoute FastPath from AVS to the Azure NetApp Files volumes virtual network for optimal performance.

- Create multiple 4-TB data stores for improved performance. The default limit is 64, but this can be changed to 256 by opening a support ticket with Azure support.

To attach the Azure NetApp Files volume to AVS, follow these steps:

1. Go to your AVS console and, under **Manage**, select **Storage (preview)**.

2. Select + **Connect Azure NetApp Files volume**:

Figure 9.24 – Connecting ANF to an AVS data store

3. On the **Connect Azure NetApp Files volume** page, select your **Subscription**, **NetApp account**, **Capacity pool**, and **volume** values that will be attached as a data store. You created these in the previous steps:

Figure 9.25 – Connecting ANF to an AVS data store

4. Under **Associated cluster**, select the **Client cluster** property to associate the Azure NetApp Files volume as a data store.

5. Under **Data store**, create a friendly name for **Datastore name**.

You will see the data store in the **Storage (preview)** tab:

Figure 9.26 – Azure NetApp Files volume in the AVS portal

As you have seen, you can increase your AVS data store using Azure NetApp Files instead of expanding your AVS nodes. This solution can be a cost saver while you're using a very high throughput disk system.

Summary

In this chapter, we looked at the storage concepts for AVS. First, we looked at vSAN, which comprises the storage from each host in the cluster. We also looked at the different storage policies available for the vSAN storage.

Fault tolerance is critical to the AVS solution, and we looked at the options available and showed you when to choose one over the other.

At some point, a customer will need to expand their AVS data store. This can be done by adding additional nodes, which can be very costly. Another option is to extend the data store using a solution such as Azure NetApp Files. This is an excellent option that will enable a customer to have a high throughput disk system, which also comes at a reduced price point.

The next chapter will look at VMware **Site Recovery Manager (SRM)**. SRM is a disaster recovery option from VMware that can be used to replicate from a primary AVS site to a secondary AVS site.

You will learn about the prerequisites needed to deploy SRM. You will also learn how to protect your primary AVS environment using SRM.

10
Working with VMware Site Recovery Manager

This chapter will look at VMWare **Site Recovery Manager (SRM)** as a **disaster recovery (DR)** option for AVS. It is crucial to deploy a solution to minimize downtime of the virtual machines in an AVS environment if there is a disaster.

Azure Site Recovery (ASR) is a native VMware DR solution that simplifies management and automation and ensures a fast and highly predictable recovery time is implemented. This will remove manual steps during a disaster and keep your business running as desired.

By the end of this chapter, you will understand what VMware SRM is and why it is crucial to have a DR solution in place for your AVS environment. We will also walk you through identifying the need for your company's DR solution.

Using the supported scenarios, you will also learn how to deploy and configure VMware SRM for your AVS environments.

Throughout this chapter, we will look at the following areas regarding VMware SRM for AVS:

- Understanding what SRM in AVS is
- Identifying your company's BCDR needs
- Installing SRM in your primary and secondary AVS environments
- Configuring site pairing for vCenter
- Connecting the SRM instances on both the protected and recovery sites

Understanding what SRM in AVS is

SRM is a DR tool created to reduce virtual machine downtime in an AVS environment in an emergency. SRM orchestrates and automates failover and failback processes to minimize downtime during emergencies. Additionally, built-in non-disruptive testing guarantees that your recovery time goals are accomplished. Overall, SRM enables quick and highly predictable recovery times while automating administration to make it simpler.

The replication technique for vSphere VMs provided by VMware is based on the hypervisor. It shields virtual machines against partial or whole site failures. Additionally, it simplifies DR protection with replication that is VM-centric and storage-independent. The ability to customize vSphere replication on a per-VM basis gives users greater control over which VMs are replicated.

> **Important note**
> SRM is not a part of the default deployment of AVS. You will need to bring your own license.

A strong **business continuity and disaster recovery** (**BCDR**) plan attempts to shield a business against downtime, financial loss, and data loss in the case of a disruptive incident. To be prepared for a disaster, you should consider several BCDR variables, just like in an on-premises VMware environment. In the following sections, we will walk you through some of these critical design considerations for AVS when using VMWare SRM as your BCDR solution.

Business continuity and disaster recovery

Any interruption in service might be inconvenient for your company and customers. Every second that your systems are down might mean lost income for your firm. Your organization may also suffer financial penalties if it fails to meet any availability agreements it has in place for the services it delivers.

BCDR plans are legal papers that businesses produce to outline the scope and measures they would take in the event of a catastrophe or large-scale outage. Each outage is evaluated on its own merits by the organization. For example, an organization may implement a BCDR strategy when a data center loses power.

Identifying your company's business continuity and disaster recovery needs

You must first assess the company's existing BCDR strategy to safeguard your organization's workloads from unanticipated events. It would be best if you determined the various recovery goals and scope for the systems that need protection. This should include the following:

- **Recovery time objective (RTO):** An RTO measures how long your company can operate after a catastrophe before it is necessary to resume regular operations to prevent unacceptably adverse outcomes brought on by a disruption in continuity.

- **Recovery point objective (RPO):** A company may back up its data every 24 hours, 12 hours, or even in real time. However, data loss is unavoidable in the event of a calamity. An RPO is a metric that measures the greatest amount of data loss that may be tolerated after a catastrophe.

VMware SRM is a BCDR solution that aids in the planning, testing, and execution of VM recoveries between a protected VMware vCenter Server site and a recovery vCenter Server site.

Supported scenarios for SRM

SRM aids in the planning, testing, and execution of VM recovery operations between protected and recovery vCenter Server sites. The following two DR scenarios are compatible with SRM and Azure VMware Solution:

- On-premises VMware to AVS private cloud DR

- Primary AVS to Secondary AVS private cloud DR

In this section, we will walk through the process of implementing DR using SRM for AVS-based VMs in a primary site replicated to a secondary site.

The architecture of a primary AVS environment to a secondary AVS environment scenario is shown in the following diagram:

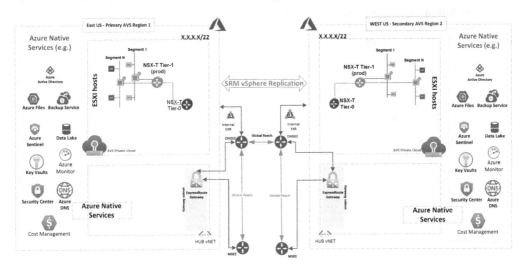

Figure 10.1 – SRM replication for primary and secondary AVS environments

The following configurations are depicted in the architecture outlined in *Figure 10.1*:

- Both Microsoft Enterprise Edge circuits are connected by ExpressRoute Global Reach

- AVS is deployed in both primary and secondary sites

BCDR support types used by SRM

SRM can be used to support the following types of recovery:

- **Disaster recovery**: When the protected AVS site goes down unexpectedly, DR while using SRM may be initiated. SRM manages the recovery process in conjunction with the replication systems to reduce data loss and system downtime.

  Individual VMs can only be secured on a host using SRM in conjunction with vSphere Replication in Azure VMware Solution.

- **Planned migration**: When both the main and secondary Azure VMware Solution sites are up and operating and completely functioning, the migration will begin. When migrating workloads in an orderly method, no data loss is envisaged when virtual machines are moved from the protected site to the recovery site.

- **Bidirectional protection**: To protect VMs in both directions, bidirectional protection employs a single set of paired SRM sites. Each site may be both a protected and a recovery site at the same time, but only for a subset of the VMs.

Throughout the rest of this chapter, we will deploy and configure SRM to protect primary and secondary AVS sites. The following tasks will be performed:

- Installing SRM in your primary and secondary AVS environments

- Installing the vSphere Replication appliance

- Configuring site pairing for vCenter

- Connecting the SRM instances on both the protected and recovery sites

- Configuring virtual machine replication

- Creating and managing protection groups for SRM

- Testing and running a recovery plan

The prerequisites are as follows:

- Provide the remote user with the VRM and SRM administrator rights in the remote AVS site
- AVS deployed in both the primary and secondary sites
- ExpressRoute Global Reach configured between both AVS sites
- SRM license (for testing purposes, you can use the evaluation version)
- SRM and vSphere Replication appliances in each site

Installing SRM in your primary and secondary AVS environments

In this section, we will walk through the process of deploying SRM in both your primary and secondary AVS environments.

> **Important note**
>
> ExpressRoute Global Reach should be configured between both the primary and secondary AVS environments.

Deploying SRM in AVS

The steps are as follows:

1. Log into the Azure portal and go to the **AVS** window.
2. Under **Manage**, select **Add-Ons**.
3. Select **Disaster recovery**. From the drop-down box, select **VMware Site Recovery Manager (SRM)**.
4. Under **Deploy SRM appliance**, select **I don't have a license key. I will use the evaluation version**. (Select **I have a license key** if you have one and then enter the license key.)
5. Select the **I agree with terms and conditions** checkbox.

6. Click the **Install** button:

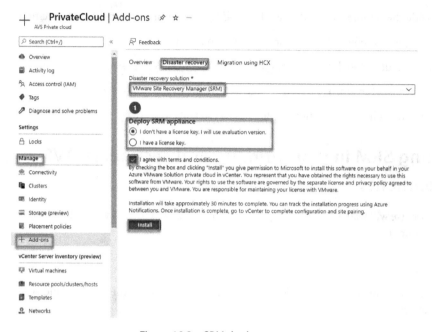

Figure 10.2 – SRM deployment

This process takes 8 to 10 minutes to complete.

Installing the vSphere Replication appliance

After the SRM appliance has been successfully installed, the vSphere Replication appliance will need to be installed. Each replication server may house up to 200 protected virtual machines. You can scale these up or down to meet your requirements.

VMware vSphere Replication is an integrated component of VMware vSphere that uses a hypervisor-based VM replication engine.

Components of the vSphere Replication appliance

vSphere Replication is an add-on for AVS that may assist in safeguarding your VMs against partial or total site failure. vSphere Replication captures any changes to the primary site VM and applies them to the VM's offline disk copies (replicas).

For protection situations, the vSphere Replication appliance has the following components:

- An SRM user interface for using vSphere Replication
- A vSphere web client and vSphere client plugin to display the health status of vSphere Replication

- The replication settings are stored in a VMware standard embedded PostgreSQL database
- A server that controls replication in vSphere Replication
- A vSphere Replication server that serves as the vSphere Replication infrastructure's heart

Follow these steps to install the add-on in AVS:

1. From the **AVS** page, select **Add-ons** from the **Manage** section.

2. Select **Disaster recovery**.

3. Under section 2, **Setup replication**, make sure **vSphere Replication** is selected from the dropdown and click the **Install** button:

Figure 10.3 – vSphere Replication appliance deployment

Please note that the number of servers in your AVS cluster will be reflected when you are installing the vSphere Replication appliance.

Configuring site pairing for vCenter

You now need to pair instances of VMware SRM on the protected and recovery sites, then establish a protection policy, to finish the process of securing your VMware vSphere virtual machines. Site pairing is the method used to connect VMware SRM instances. A Windows client virtual machine, or jumpbox, installed on the virtual network with access to both AVS private clouds may be used.

Once installed, verify that both the SRM and the vSphere Replication appliances are installed.

Configuring site pairing in vCenter

Follow these steps to configure site pairing:

1. Ensure that the jump box device has connectivity to both the primary and secondary AVS environments.

2. From the jump box VM, open a browser and connect to AVS vCenter Server and NSX-T Manager using the credentials that you can retrieve from the **Identity** pane in your **AVS** page in Azure.

3. In the vSphere web client, click on **vSphere Client | Site Recovery**:

Figure 10.4 – vSphere Site Recovery

4. In the **Site Recovery** window, verify that an **OK** status displays for both **vSphere Replication** and **Site Recovery Manager**. Then, click on **OPEN Site Recovery**:

Figure 10.5 – The Site Recovery window

5. Click on the **NEW SITE PAIR** button:

Figure 10.6 – The NEW SITE PAIR window

6. Click on the first site that is shown on the list. Click **NEXT**.

7. On the **Peer vCenter Server** window, enter the required information for the Platform Services Controller for the SRM server on the secondary AVS environment, as detailed in the following screenshot. Click on **FIND VCENTER SERVER INSTANCES**:

Figure 10.7 – Peer vCenter Server

8. You will be prompted with a security error for the default certificate that is on the destination vCenter server. This happens because the evaluation version of SRM is being used. Click on **CONNECT**:

Figure 10.8 – Site pair security alert

9. In the **Peer vCenter Server** window, click on the radio button next to the other vCenter server that you want to pair.

10. Select **CONNECT** to accept the certificates for the remote VMware SRM and the remote vCenter Server (again).

11. Select **CONNECT** to accept the certificates for the local VMware SRM and the local vCenter Server.

12. Click **NEXT**:

Figure 10.9 – Secondary vCenter Server

13. On the **Ready to complete** window, take note that vCenter Server instances are displayed. Also, look at **vSphere Replication** and **Site Recovery Manager**.

14. Click on **FINISH**:

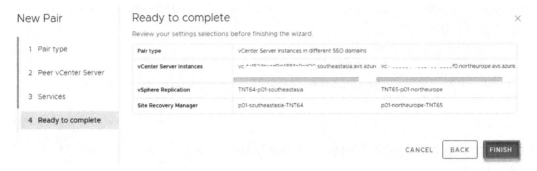

Figure 10.10 – Completing site pairing

This process takes a few minutes to complete. Once the site pairing has been completed, you will be able to view the details in the **Site Recovery** window:

Figure 10.11 – Newly created site pair

We now have the newly created site pair. Now, let's connect the SRM instances on both the protected and recovery sites.

Connecting the SRM instances on both the protected and recovery sites

Following the successful configuration, you must link the VMware SRM instances on both the protected and recovery sites.

> **Important note**
> The following ports should be open to provide cloud-to-cloud recovery: 80, 443, 902, 1433, 1521, 1526, 5480, 8123, 9086, 31031, 32032, 8043, and 10000-10010.

Configuring mapping between both the primary and secondary SRM sites

Before you can begin safeguarding the VMs, you must first map the items on the protected site to their counterparts on the recovery site. You may map the following items to verify the functioning of replicated VMs:

- Networks
- VM folders
- Compute resources
- Storage policy mappings

> **Important note**
> To allow bidirectional protection, reverse mappings may be configured to map items on the recovery site to their equivalent objects on the protected site. You will be required to log into both sites to configure resource mapping.

The steps to create a mapping in SRM are as follows:

1. Log in to vCenter using your CloudAdmin credentials.
2. Go to **Site Recovery**, click on **Menu**, and select the new site pair that you created earlier:

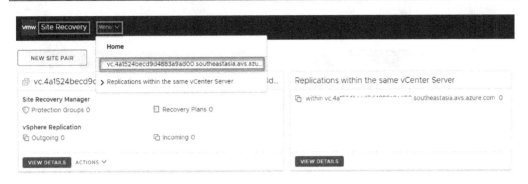

Figure 10.12 – Newly created site pair

3. Click on **LOGIN** to authenticate to the first site:

Figure 10.13 – Authenticate to SRM site

4. In the **Log In Site** window, enter the cloudadmin credentials for the first site and click on **LOG IN**:

Figure 10.14 – cloudadmin login

Once you have authenticated, you will be able to view the Site Recovery summary:

Figure 10.15 – Site Recovery summary

Creating a new network mapping

The steps are as follows:

1. To create a new network mapping, under **Site Pair**, click on **Network Mappings** and click **NEW**:

Figure 10.16 – New network mapping

2. Select *automatic or manual* mapping from the **Creation mode** window. For this exercise, we will be using the **Automatically prepare mappings for networks with matching names** option. Click **NEXT**:

Figure 10.17 – New network mapping creation mode

3. Expand **SDDC-Datacenter** on both sites.

4. Click the checkbox next to **Static-VM** on both sites and click **ADD MAPPINGS**:

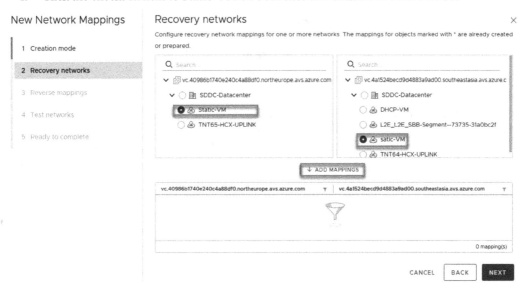

Figure 10.18 – Network mapping selection

5.   Click **OK** on the **Discovered Mappings** pop-up window.

6.   You will now see the network mapping, as shown in the following screenshot. Click **NEXT**:

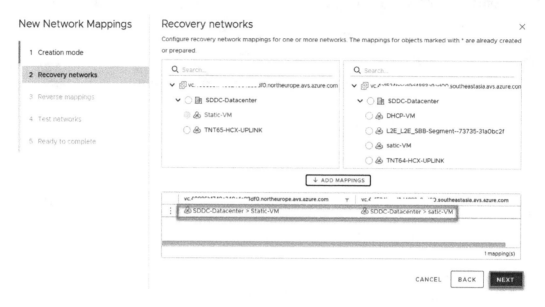

Figure 10.19 – New network mapping created

7.   (Optional) Select the checkbox on the **Reverse mappings** window. Click **NEXT**:

Important note

Selecting this option generates equivalent mappings from the secondary site to the primary site. Reverse mappings are required for bidirectional protection and reprotection procedures. This option is not available if two or more mappings have the same target on the remote site.

Figure 10.20 – Reverse mappings

8. (Optional) In the **Test networks** window, click **CHANGE** and, on the **Edit Test Network** page, select the network that you will use when testing recovery plans.

> **Important note**
>
> SRM may be configured to establish an isolated network on the recovery site for testing recovery plans. Creating an isolated test network enables the test to complete without introducing additional traffic to the recovery site's production network.

9. For this exercise, we will deploy the test network. Click on **NEXT**:

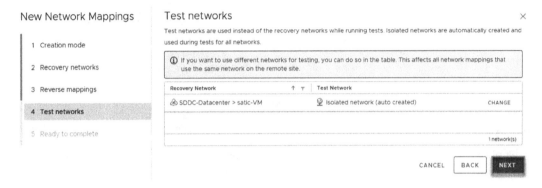

Figure 10.21 – Creating a test network

10. In the **Ready to complete** window, click on **FINISH**:

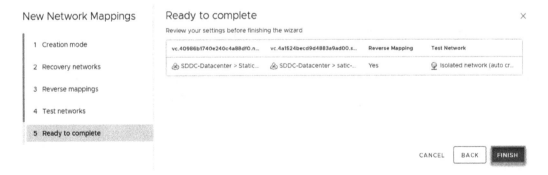

Figure 10.22 – Finish network mappings

11. On the **Site Recovery** page, click on **Network Mappings**. You will now see the new mapping that was created earlier:

Figure 10.23 – Network Mappings

You have now successfully created a network mapping for Site Recovery.

Repeat *steps 1* through *10* to create mappings for **VM folders**, **Compute resources**, and **Storage policy**.

By defining mappings, you can guarantee that the VMs have access to all the resources that are accessible in the recovery site. If you don't make the necessary mappings, you'll have to modify these parameters for each VM manually.

A crucial part of VMWare SRM is ensuring that the virtual machines running your company's applications are always protected. In the next section, we will walk you through how to configure replication for those virtual machines.

Configuring virtual machine replication

You must set up replication on the VM you want to safeguard. When configuring replication settings, you may specify several point-in-time instances that will be converted into snapshots after recovery.

Follow these steps to configure virtual machine replication:

1. Click on **Site Recovery** in the vSphere Client, and then select **Open Site Recovery**.

2. Click on **Menu**. Then, click on the site pair that was created earlier.

3. Select the **Replication** tab and select **New**.

4. On the **Target site** page, make sure that a target site is selected.

5. Click **NEXT** to continue with the default option of **Auto-assign vSphere Replication Server**:

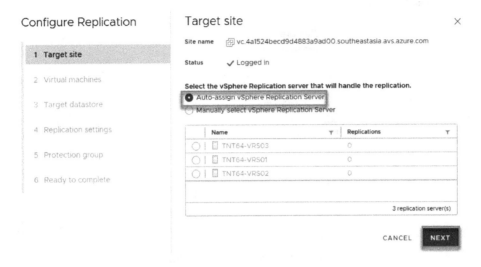

Figure 10.24 – VM replication target site

6. On the **Virtual machines** page, select the virtual machines that you want to protect.

7. Click **NEXT**:

Figure 10.25 – Virtual machines selected for protection

8.   On the **Target datastore** page, configure the **Disk format** and **VM storage policy** options for the protected VMs and click **NEXT**:

Figure 10.26 – Target datastore options

9.   On the **Replication settings** page, select a recovery point objective that meets your organization's needs. You also have the option to enable point-in-time instances, network compression, and encryption of the data. Select **NEXT**:

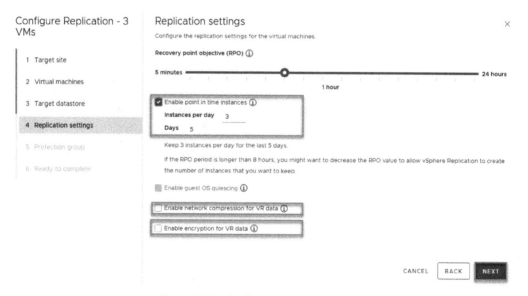

Figure 10.27 – Replication settings options

10. On the **Protection group** page, you have the option to add the virtual machines to an existing protection group, create a new protection group, or not add them to a protection group now. We will create a protection group in the next section, so select **Do not add to protection group now** and click **NEXT**:

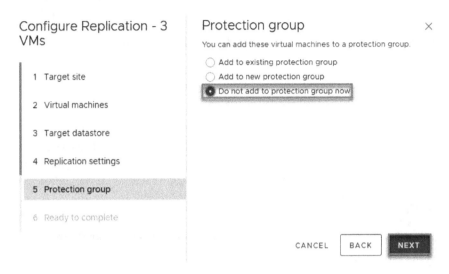

Figure 10.28 – Protection group options

11. On the **Ready to complete** page, click **FINISH**.

You will now be able to see the VM replication that was just created in Site Recovery under the **Replications** tab:

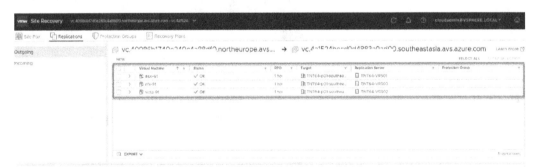

Figure 10.29 – VM Replications view

With that, you have created a VM replication plan. Next, you will need to create a protection group for the VMs in the replication plan.

Creating and managing protection groups for SRM

Multiple VMs may be included in a protection group that VMware SRM will protect. The recovery plan, which describes how VMware SRM recovers the VM housed in the protection group, includes protection groups.

The VMs in the replication plan must be assigned to a resource pool, folder, and network that already exist on the recovery site. These parameters may be customized for each VM in the protection groups separately or inside the inventory mappings.

To apply the inventory mappings to each VM in the group, VMware SRM first generates placeholder VMs on the recovery site. Then, as per the recovery point goal that you specified when you set up vSphere Replication on the VM, vSphere Replication synchronizes the disk files of the replication target VM.

Follow these steps to create a vSphere Replication protection group:

1. Click on **Site Recovery** in the vSphere client, and then select **Open Site Recovery**.

2. Select the **Protection Groups** tab and click on **NEW**.

3. On the **Name and Direction** page, enter a unique name and description, select a direction, and then click on **NEXT**:

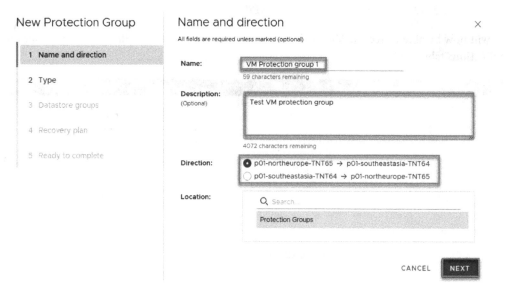

Figure 10.30 – Protection group name and direction

4. On the **Type** page, select **Individual VMs (vSphere Replication)**, and then select **NEXT**:

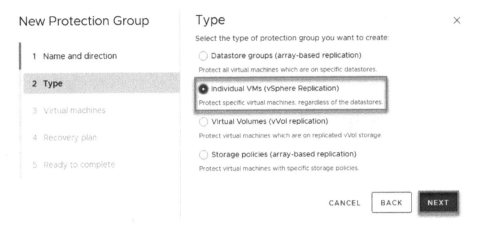

Figure 10.31 – Protection group type

5. On the **Virtual machines** page, select the VMs that you want to add to the protection group. (Only VMs that are not a part of a protection group will be listed.) Click **NEXT**:

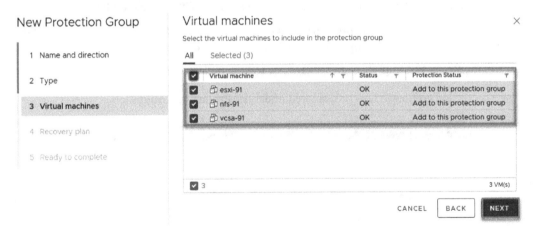

Figure 10.32 – Protection group VMs

6. On the **Recovery plan** page, you may add a protection group to a recovery plan by choosing one of the following choices:

- **Add to existing recovery plan**
- **Add to new recovery plan**
- **Do not add to a recovery plan now**

7. Select **Add to new recovery plan**. Enter a unique name for the recovery plan and click **NEXT**:

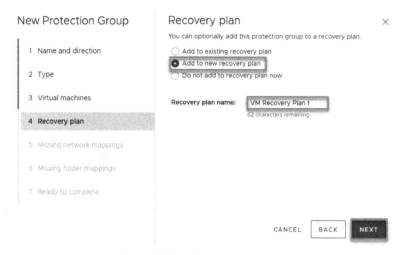

Figure 10.33 – New recovery plan

8. On the **Missing network mappings** page, select one or more networks from each site to add to the recovery network mapping. Click on **ADD MAPPINGS**, then click **NEXT**.

9. On the **Missing folder mappings** page, select one or more folder options from each site to add to the recover folder mappings. Click **ADD MAPPING** and then click **NEXT**.

10. On the **Ready to complete** page, review your settings and click **FINISH**.

The new protection group will now be displayed under the **Protection Groups** tab:

Figure 10.34 – New protection group

Now that we have created a protection group and a recovery plan, we will test and run the recovery plan.

Testing and running a recovery plan

During a test of the recovery plan, the source VM continues to function in the primary site, and a replica of that VM is produced in the recovery site in the test network.

Follow these steps to test the recovery plan:

1. In the **Site Recovery** window, click on the **Recovery Plans** tab.

2. Select the radio button next to the recovery plan and click on **TEST**:

Figure 10.35 – Testing a recovery plan

3. On the **Confirmation options** page, make sure that you select the **Replicate recent changes to recovery site** checkbox, and then select **NEXT**:

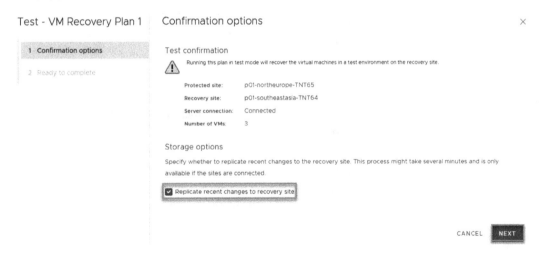

Figure 10.36 – Testing a recovery plan confirmation option

4.   Click **FINISH** on the **Ready to complete** page after reviewing your options.

5.   After a successful test, make sure you clean up your environment before running the recovery plan:

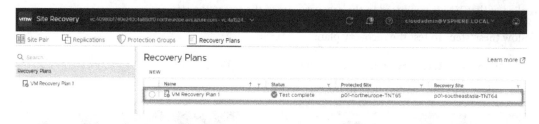

Figure 10.37 – Successful test recovery plan

Running a recovery plan

When you choose to run the recovery plan, a scheduled migration is initiated. Planned migration is the transfer of virtual machines from the protected site to the recovery site in an organized manner. Planned migration avoids data loss during the orderly relocation of workloads. For the intended migration to be successful, both sites must be fully up and functioning. Then, if the protected site fails, DR may be initiated to restore the failed VMs. VMware SRM orchestrates the recovery process in conjunction with replication technologies to reduce data loss and system downtime.

Follow these steps to run a recovery plan:

1.   In the **Site Recovery** window, click on the **Recovery Plans** tab.

2.   Select the radio button next to the recovery plan and click on **RUN**.

3.   On the **Confirmation options** page, select the **I understand that this process will permanently alter the virtual machines and infrastructure of both the protected and recovery datacenters** checkbox.

4. Under **Recovery type**, select **Planned migration** and click **NEXT**:

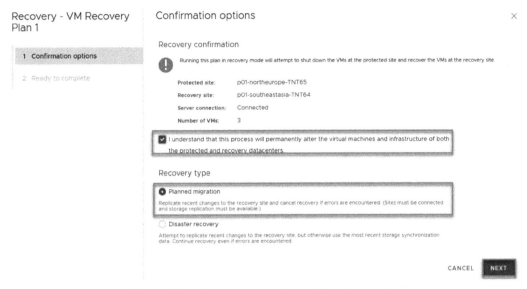

Figure 10.38 – Run a recovery plan confirmation option

5. On the **Ready to complete** page, click **FINISH** after reviewing your options.

With that, you have implemented, configured, and tested a recovery plan for your AVS environments.

Summary

In this chapter, VMWare SRM was the focus. You have seen the importance of having a DR solution for your AVS environments and that SRM is a key solution that is being utilized by many customers.

Then, you learned how to deploy and configure SRM in an AVS environment. You also learned how the different components of SRM are configured.

Finally, you learned how to test a recovery plan and how to run a recovery plan in case a need is there for it.

In the next chapter, you will learn how to manage an AVS environment. You will learn about the different responsibilities and toolsets that can be used to manage your environment.

Part 4: Governance and Management for AVS

The purpose of this part is to discuss the governance and management methods used across AVS, as well as to walk through their best practices.

This part comprises the following chapters:

- *Chapter 11, Managing an Azure VMware Solution Environment*
- *Chapter 12, Leveraging Governance for Azure VMware Solution*
- *Chapter 13, Summary of Azure VMware Solution, Roadmap, and Best Practices*

11
Managing an Azure VMware Solution Environment

AVS is a VMware-validated solution that undergoes continuous validation and testing for vSphere advancements and updates. Microsoft administers and supports a customer's private cloud infrastructure and software. Microsoft assumes this duty, allowing you to concentrate on creating and executing workloads inside your private cloud and Azure-native resources. Regular updates of the AVS private cloud and VMware software guarantee that your deployed private cloud has the most recent security, stability, and feature sets.

Microsoft routinely displays the shared responsibility matrix for IaaS, PaaS, and SaaS-based solutions. A shared responsibility matrix is also available for AVS. As can be seen, Microsoft abstracts a significant portion of continuous maintenance, security, and administration, putting your organization in control of what matters most, such as guest OS provisioning and virtual machines. You may also consider your life cycle process and configuration management strategies that can be implemented on Azure. Using this architecture eliminates some operational obligations, as Microsoft assumes more responsibility for the AVS infrastructure.

Microsoft is responsible for the underlying infrastructure, which includes the security and patching of the AVS nodes when you install AVS in Azure. This responsibility changes the traditional process usually used in an on-premises data center. This change is because AVS is an Azure Managed Service. This change also allows IT staff to focus on driving change within their mission-critical applications and workloads. It facilitates a digital revolution that transcends AVS.

The following table shows the shared responsibilities of Microsoft and customers alike.

	Deployment	Life Cycle	Configuration
Physical infrastructure	Microsoft	Microsoft	Microsoft
Physical security	Microsoft	Microsoft	Microsoft
Azure/AVS portal	Microsoft	Microsoft	Microsoft
Hardware failure	Microsoft	Microsoft	Microsoft
ESXi host	Microsoft	Microsoft	Microsoft
Host patching	Microsoft	Microsoft	Microsoft
NSX-T	Microsoft	Microsoft	Customer
Identity management	Microsoft	Microsoft	Customer
vCenter	Microsoft	Microsoft	Customer
vSAN	Microsoft	Microsoft	Customer
Virtual machines	Customer	Customer	Customer
Guest OS	Customer	Customer	Customer
Applications	Customer	Customer	Customer

■ Microsoft
■ Customer

Table 11.1 – AVS shared responsibility

Throughout this chapter, we will be focusing on the following key areas:

- AVS business alignment
- Managing and monitoring your AVS environment

- Configuring Azure Alerts in AVS

- VMware Syslog configuration for AVS

AVS business alignment

IT assets (applications, virtual machines, VM hosts, disks, servers, devices, and data sources) are managed by the IT department in on-premises environments to support workload operations. IT management provides processes to help support business operations by minimizing disruptions to those assets. We have seen many times when the IT department would like to redo these processes for a more stringent operation but is limited by the possibility of downtime or taking systems offline. When a company migrates to the cloud, management and operations shift slightly, creating an opportunity for tighter business alignment for an even more robust process and less downtime.

Creating business alignment begins with term alignment. IT management has accumulated buzzwords or highly technical terms, just like most engineering professions. Such words can perplex business stakeholders and complicate mapping management services to business values.

Some terminology, as we know them on-premises, changes when you migrate to Azure. The same is for AVS. In AVS, there are no VLANs. However, segments are used instead.

Fortunately for us as IT professionals, the process of building a cloud adoption strategy and cloud adoption plan offers the perfect chance to remap these terms. This approach also allows us to rethink operational management commitments in collaboration with the company's IT goals for the future.

Managing and monitoring your AVS environment

Now that you have deployed your AVS environment, it is time to focus on the management and monitoring aspect to ensure your environment is operating at the highest level to maintain your company's **service-level agreement (SLA)**.

The success of your AVS environment is dependent on proper management and monitoring. It is critical to understand the shared responsibility matrix as you plan your management and monitoring environment for AVS. *Table 11.1* show the areas of responsibility for both Microsoft and the customers. Microsoft handles the ongoing maintenance, security, and management of cloud resources, leaving your company to focus on what matters most, such as guest OS provisioning, applications, and virtual machines.

AVS platform monitoring and management

The following recommendations are to help you monitor and manage your AVS platform.

Azure native toolset recommendations:

- A local identity provider is used by AVS. After deployment, configure AVS using a single administrative user account. Integrating AVS with Active Directory Domain Services allows you to track user actions. You will be able to integrate AVS with your AD environment by using the **Run command** area in the AVS portal and then selecting the **New-LDAPSIdentitySource** option:

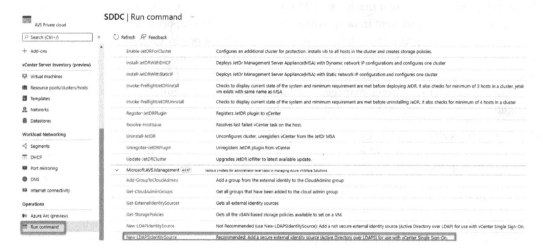

Figure 11.1 – AD integration using New-LDAPSIdentitySource

- The Activity Log keeps track of all actions taken within Azure. These activities include key and credential listing, as well as creation, updating, and deletion. When someone accesses the **Identity** tab of the AVS portal or makes a programmatic request for cloud admin credentials, for instance, AVS will issue a **List PrivateClouds AdminCredentials**. Notifications can be configured to be sent when certain activities are logged using alert rules.

- To maintain availability and performance, vSAN storage must be managed properly due to its limited resource. Only use vSAN storage for workloads on guest VMs. Learn the concepts of storage for AVS (see *Chapter 9*).

- For the KPIs that matter most to your operations teams, create alerts and dashboards.

- Configure **Azure Service Health** to provide notifications for service problems, scheduled maintenance, and other occurrences that could affect AVS. These notifications are sent to **Action Groups**, which can be used to send voice calls, SMSs, emails, and push notifications to addresses of your choice. Actions can also be used to trigger Azure and third-party systems, such as Logic Apps, Automation Runbooks, Event Hubs, and Webhooks.

- Azure Monitor Metrics can be used to monitor the baseline performance of your AVS infrastructure. These metrics can be queried and filtered through the Azure portal, via the REST API, or by directing them to Log Analytics, Azure Storage, Event Hubs, or Partner Integrations.

- Set up the following alerts in Azure Monitor to send notifications when the cluster's disk, CPU, or RAM usage is approaching a critical level:

Metric	Alert
CPU – Percentage CPU %	>80% warning
Memory – Average Memory Usage (%)	>80% warning
Disk – Percentage Datastore Disk Used %	>75% critical
Disk – Percentage Datastore Disk Used (%)	>70% warning

Table 11.2 – Metrics alert details

- Both Azure Monitor Notifications and Azure Service Health alerts can be automated.

- AVS needs 25% of the available slack space on vSAN to meet the SLA.

- AVS requires the number of failures to tolerate = 1 for clusters with 3 to 5 hosts, and the number of failures to tolerate = 2 for clusters with 6 to 16 hosts, to comply with SLA requirements.

- You can use Connection Monitor in a hybrid environment to monitor communication between on-premises and your Azure resources.

In a hybrid environment, you can use Connection Monitor to monitor communication between on-premises and Azure resources.

I will walk you through creating an alert for AVS using the Azure portal in the *Configuring Azure Alerts for AVS* section later in this chapter.

VMware toolset recommendations

Some of the VMWare toolset recommendations are as follows:

- With the aid of the diagnostic settings found in the AVS portal under **Monitoring**, vCenter logs can be delivered to **Storage Accounts** or **Event Hubs**. Log settings can only be configured via the Private Cloud resource and don't explicitly state this has to be done through the diagnostic settings.

  Log settings aren't directly configurable within vCenter, only via the Private Cloud resource in Azure.

- While Microsoft monitors the health of vSAN, vCenter may be used to query and monitor vSAN's performance. Through vCenter, performance metrics may be viewed from the standpoint of a VM or a backend, displaying average latency, IOPS, throughput, and outstanding I/O.

- To give you a better understanding of the AVS platform, think about VMware products such as vRealize Operations Manager and vRealize Network Insights. Customers can view NSX-T distributed firewall monitoring information such as vCenter events and flow logs.

- At the moment, vRealize Log Insight for AVS supports pull logging. Only events, tasks, and alarms can be captured. It is not currently possible to push unstructured data from hosts via Syslog to vRealize. SNMP traps cannot be used.

VM workload management recommendations

Some of the VM workload management recommendations are as follows:

- Without deploying Azure Arc for Servers onto AVS VMs, they won't appear in the Azure portal. An agent-based method of managing and monitoring virtual machines from the Azure control plane is possible with Azure Arc for Servers. Azure Policy guest setups, Microsoft Defender server security, and deploying the Azure Monitor agent to the guest VMs are all options.

  The following screenshot shows that no VMs are visible in the AVS portal because Azure Arc is not deployed:

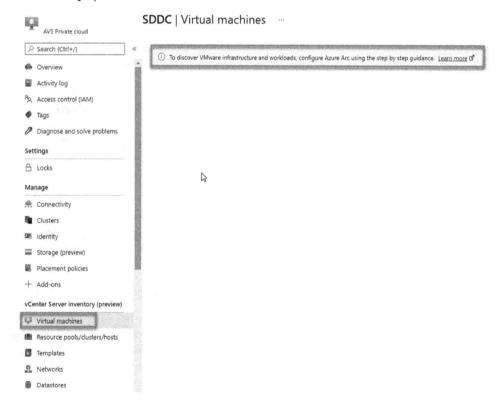

Figure 11.2 – Azure Arc not deployed for AVS

- By default, virtual machines within AVS are handled in the same way as VMware VMs located on-premises. You can keep utilizing the current agents in AVS for VM-level monitoring.

- Thick provisioning is used in the default storage policy. Consider adopting thin provisioning for VMs for effective vSAN capacity use. Disk setup can differ for every VM. Depending on the needs of the workload, a VM may contain thick, thin, or both types of disks.

The following is a list of tools that you can utilize to help with monitoring your AVS environment:

Integration Tool	Description
Azure Monitor	An integrated monitoring tool for gathering, examining, and responding to telemetry from on-premises and cloud environments.
Log Analytics	The primary tool for aggregating, querying, and interactively analyzing logs generated by Azure resources.
Azure Update Management	Manages operating system updates for Windows and Linux machines on-premises and in cloud environments.
Microsoft Sentinel	A solution for managing security-related information and events in the cloud. In both on-premises and cloud contexts, this Azure resource offers security analytics, alert detection, and automated threat response.
Microsoft Defender for Cloud	By offering sophisticated threat prevention across hybrid and Azure resources, a unified infrastructure security management system improves the security posture.

Table 11.2 – Azure monitoring integration tools

Configuring Azure Alerts for AVS

In this section, you'll learn how to configure Azure Action Groups in Microsoft Azure Alerts to get notifications of triggered events that you designate. Additionally, you'll discover how to use Azure Monitor Metrics to better understand your private cloud powered by the AVS.

> **Important note**
> The Account Administrator, Service Administrator (Classic Permission), Co-Admins (Classic Permission), and Owners (RBAC Role) of the subscription(s) containing AVS private clouds are automatically notified of incidents affecting an AVS host's availability and their corresponding restoration.

Azure Monitor supported metrics and activities

Through **Azure Monitor Metrics**, the following metrics are visible:

Signal Name	Signal Type	Monitor Service
Datastore Disk Used	Metric	Platform
Average Memory Usage	Metric	Platform
Average Total Memory	Metric	Platform
Average Memory Overhead	Metric	Platform
Average Effective Memory	Metric	Platform
Percentage CPU	Metric	Platform
Percentage Datastore Disk Used	Metric	Platform
Datastore Disk Total Capacity	Metric	Platform
Delete a PrivateCloud. (Microsoft.AVS/privateClouds)	Activity Log	Administrative
Create or update a PrivateCloud. (Microsoft.AVS/privateClouds)	Activity Log	Administrative
Register Microsoft.AVS resource provider. (Microsoft.AVS/privateClouds)	Activity Log	Administrative
All administrative operations	Activity Log	Administrative

Table 11.3 – Azure Monitor supported metrics, signal types, and service types

Configuring an alert rule for AVS

In the following steps, I will walk you through how to create an alert rule for AVS:

1. From your AVS private cloud, click on **Alerts** under **Monitoring** and then click on **Create alert rule**:

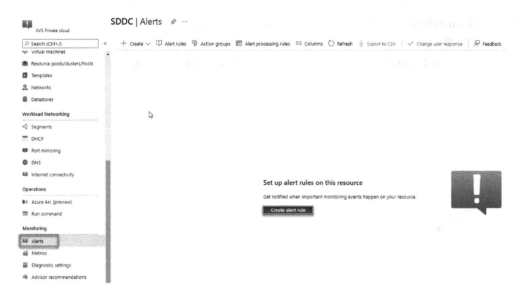

Figure 11.3 – Creating a new alert rule

You will see a new configuration screen, where you will do the following:

- Define the scope

- Configure a condition

- Set up an action group

- Define the alert rule details:

Figure 11.4 – Alert rule configuration page

2. Click on **Scope** and select the target resource you want to monitor. By default, the AVS environment from where you opened the Alerts will be selected.

3. Click on the **Conditions** tab and select **Add condition**. Select the signal you want to create for the alert rule:

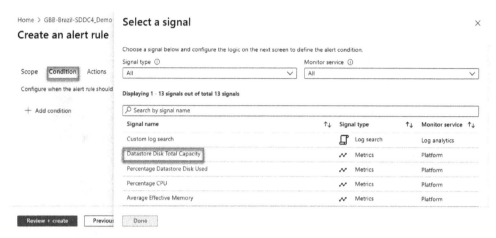

Figure 11.5 – The selected signal to be monitored

4. Define which logic will trigger the alert and then select **Done**.

    In the following example, the **Threshold** value is set to **75**, and the **Unit** value is set to **TB**:

Figure 11.6 – Alert condition configuration

5.   The new condition that you created will be displayed under the **Condition** tab:

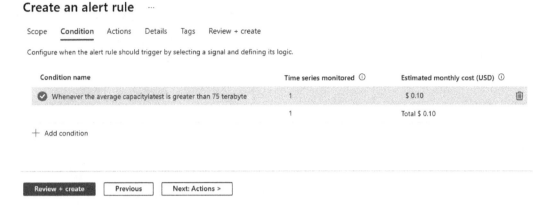

Figure 11.7 – Newly created condition

6.   Click on the **Actions** tab and select + **Create action group**. The notification's delivery method and intended audience are specified by the action group. Email, SMS, Azure Mobile App Push Notification, and voicemail are all acceptable methods of notification:

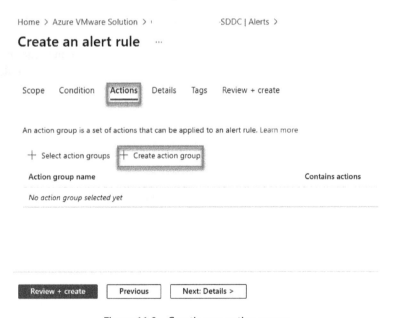

Figure 11.8 – Creating an action group

7. On the **Basics** tab, enter a name for the action group and a display name.

8. Click on the **Notifications** tab and select a **Notification Type** and **Name**. Click on **OK**. This example is based on SMS notification:

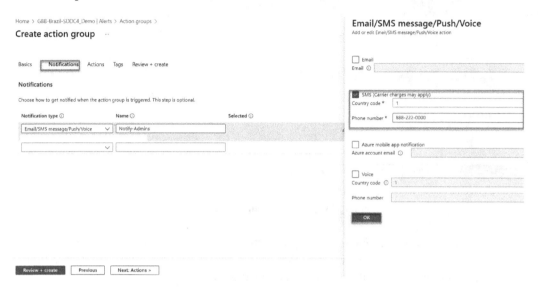

Figure 11.9 – Configuring the notification

9. After the notifications have been configured, an email or SMS will be sent to the email address or phone number that was used. In this example, SMS was configured. The following is a verification SMS that was received:

Figure 11.10 – SMS notification

10. Click on the **Actions** tab. You can choose an action to perform when the action group is triggered. This step is optional. The following screenshot shows the different action types that can be used. We won't be configuring any actions in this example:

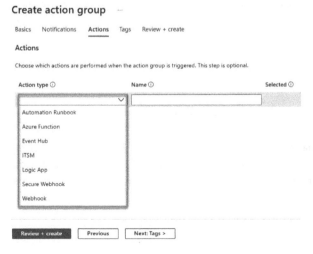

Figure 11.11 – Action type options

11. Select the **Review + create** tab and review the summary page before you click **Create**.

12. The **Alerts** menu will display an alert once the metric that was configured has been reached. Click on the **Alert rules** tab to view the alert that you created earlier:

Figure 11.12 – Alert menu

The alert rule is now displayed, and you will be able to see the configuration options that were used to create the rule. The alert rule needs to be enabled for it to trigger notifications:

Figure 11.13 – Alert rules

By now, you should have a very good understanding of the importance of monitoring your AVS environment, the different metrics to monitor, and how to configure monitoring. In the next section, we will look at different metrics that you can work with in your AVS environment.

Metric options for AVS

To see the different metric options for AVS, click on **Metrics** under the **Monitoring** section of your AVS portal. Select the metric that you want from the drop-down list:

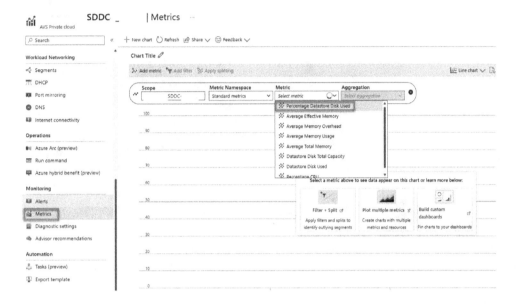

Figure 11.14 – Metric options for AVS

You have the option to change the diagram parameters, such as **Time range** or **Time granularity**:

Figure 11.15 – Metric options for AVS datastore

You can also change the chart type by selecting an option from the drop-down list:

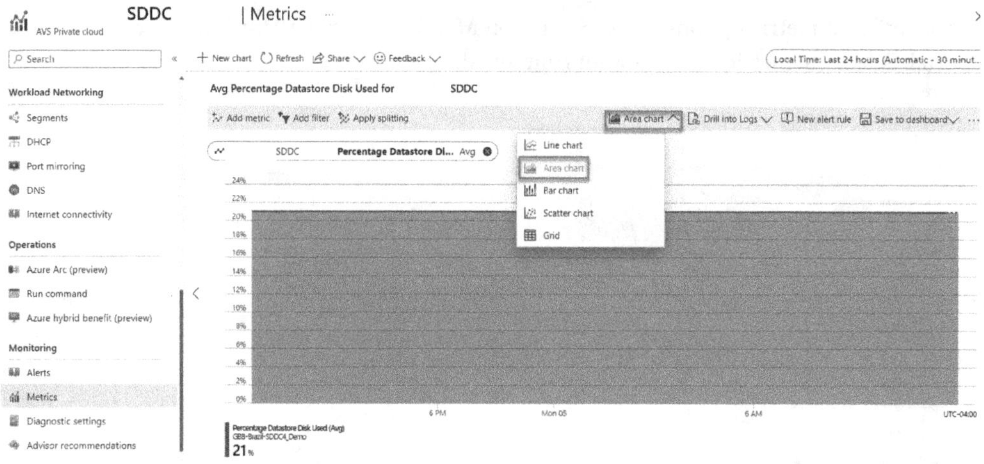

Figure 11.16 – Metric chart type

As you have seen, proper administration and monitoring are necessary for an AVS environment to be successful. You can use native Azure, VMWare, and other third-party tools to manage and monitor your AVS environment.

In the next section, you will learn how to use the VMware syslogs for AVS as a part of your monitoring routine for your AVS environment.

VMware Syslogs configuration for AVS

Diagnostic settings are used to configure the streaming export of platform logs and metrics for a resource to a specified destination. You may configure up to five diagnostic settings to deliver various logs and data to different locations.

In this section, you'll set up a diagnostic setting for your AVS environment to collect VMware syslogs. You will save these syslogs to a blob storage account so that you can look at the vCenter Server logs and analyze them for diagnostic purposes.

Prerequisites

The prerequisites are as follows:

- An AVS environment with access to the vCenter and NSX-T Manager interfaces
- An Azure storage account to save the logs to

The following logs are contained in the VMware syslogs:

- NSX-T Data Center Distributed Firewall logs

- NSX-T Manager logs

- NSX-T Data Center Gateway Firewall logs

- ESXi logs

- vCenter Server logs

- NSX-T Data Center Edge Appliance logs

Diagnostic settings configuration

The steps are as follows:

1. From your AVS portal, click on **Diagnostic settings** under **Monitoring**. Click on + **Add diagnostic setting**:

Figure 11.17 – Adding a diagnostic setting

2. On the **Diagnostic setting** page, enter a name for your **Diagnostic setting**. Select **AllMetrics** and **vmwaresyslog**. Under the **Destination details** section, select **Archive to a storage account**. Validate your **Subscription** and a **Storage account** to send the logs to. Click **Save**:

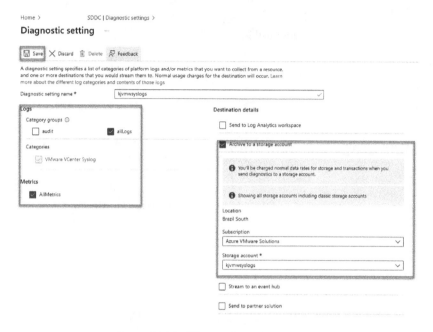

Figure 11.18 – Diagnostic setting options

3. Go to the storage account that you selected for the logs to be saved to. Click on **Containers** and verify that the **insights-metrics-pt1m** container has been created. Click on it:

Figure 11.19 – Insights-metrics-pt1m in the storage container

4. Browse through the **insights-metrics-pt1m** container to locate and download the JSON file so that you can view the logs:

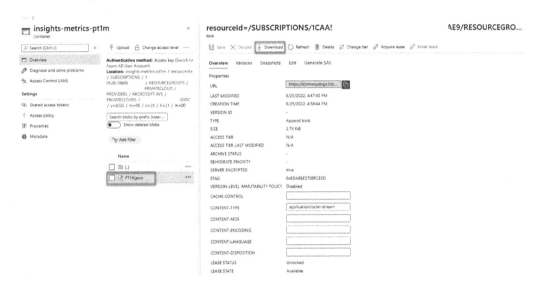

Figure 11.20– Insights-metrics-pt1m in the storage container with a JSON file

You also have the option to send the VMware syslogs to Microsoft Azure Event Hubs or a Log Analytics workspace.

Summary

In this chapter, we covered the topic of managing and monitoring your AVS environment. A management and monitoring policy must be in place to help you see and understand what is happening to your AVS environment at any given time.

You also learned how to configure monitoring and alerting, depending on the severity. Team members are notified, and the necessary actions can be taken to prevent any outages in your environment.

In the next chapter, we will look at leveraging governance for your AVS environment. You will learn about the security and compliance tools available to help you keep your environment in compliance.

We will also look at integrating some of the Azure-native services that will help with your compliance and security endeavors.

12
Leveraging Governance for Azure VMware Solution

AVS is a VMware-powered Azure first-party solution that provides vSphere clusters in a single-tenant private cloud environment. Users and apps may access it via on-premises vSphere solutions as well as Azure-based environments or resources. In Azure, the VMware technology stack uses a highly secure collection of computing, storage, and networking technologies.

An ExpressRoute circuit is not required but highly recommended to connect to Azure Cloud Services through a dedicated private and redundant Layer 3 network fiber connection with bandwidth up to 100 Gbps. You can connect your AVS environment to your Azure-native environment to consume the other Azure-native services and solutions.

All provisioned private clouds include vCenter Server, ESXi, vSAN, and NSX-T Data Center, allowing you to migrate your workloads from on-premises vSphere infrastructures, deploy new **virtual machines (VMs)**, and consume Azure services.

VMware vSphere clusters are built on top of hyper-converged, bare-metal equipment that "shares nothing." The AVS cluster design is dedicated and isolated, which means that no other tenant's networking, storage, or compute is shared. Microsoft manages the VMware vSphere clusters in Azure to fulfill performance, availability, security, and compliance needs at scale, while also offering unified management, networking, and operational controls.

Because AVS runs hybrid workloads across on-premises vSphere and private clouds, offering a single pane of glass for progressively implementing needed governance and operational management controls is the optimal way.

We will be looking at the following topics in this chapter, which will help you leverage governance for your AVS environment:

- A unified security and compliance approach
- Integrating Azure-native tools/services with AVS

- Security for your AVS environment

- VM and guest application security

- Compliance

- Governance

- Azure-native solutions integration

A unified security and compliance approach

With a shared operating framework, you can run, manage, and protect your applications across your AVS deployments. You can also use your existing VMware solution tools, such as VMware vCenter Server, vSAN, and NSX-T Manager, in conjunction with Azure's scalability, performance, and innovation. AVS should leverage vSphere role-based access control for increased protection in terms of access and security. vSphere SSO LDAP features may be integrated with Azure Active Directory.

You can evaluate and manage risk tolerance by identifying high-risk business sectors, translating risk vectors into controlling corporate policies, and extending governance rules across the Cost Management, Security Baseline, Identity Baseline, Resource Consistency, and Deployment Acceleration disciplines.

The following tables list several different Azure-native tools that you should utilize to govern your AVS environment:

Security

Azure Active Directory Domain Services	Sentinel	Security Baseline	Security Center	Azure Defender	Azure Role-Based Access Control
Azure Active Directory Group	Secure Score	Key Vault	Domain Controller	Firewall manager	Identity and Access

Governance

Azure Active Directory Privileged Identity Management	Azure Monitor	Usage and Quota	Automation Account	Backup Center
Extended Security Updates	Security Alert	Log Analytics Workspace	Identity Governance	Role (Azure AD)

Compliance

Azure Arc machine	Cost Management	Tags	Compliance	Policy	vRealize Log Insight	vRealize Network Insight

Table 12.1 – Azure-native governance tools

We will take a closer look at the aforementioned tools later in this chapter.

Integrating Azure-native tools/services with AVS

By integrating Azure-native services into your AVS environment, your workloads will benefit from unified operations best practices for governance boundaries.

Some of the available Azure-native services and their descriptions are as follows:

- **Native Azure integration**: Connect to Azure services endpoints. Deploy Azure AD as the SSO identity source for VMware vCenter Server, for example.

- **Unified VM management**: A single pane of glass for managing vSphere-based and Azure-native VMs with standardized identities, access control, and monitoring.

- **Single point of support**: Microsoft develops, manages, and supports AVS. Microsoft serves as the only point of contact and arranges support with VMware as needed. This is very important as it prevents customers from opening support tickets with multiple vendors.

- **Azure Hybrid Use Benefits (AHUB):** Optimizes the use of current Windows Server and SQL Server licenses. AHUB and Azure Reserved VM Instances provide savings of up to 80%. Microsoft is the only cloud vendor that allows customers to use their existing Windows Server and SQL Server licenses in the cloud from on-premises.

- **Unified licensing and consumption**: Avoid needless complexity and feel confident managing a single vendor for resource use and licensing.

The following is a high-level architectural overview of how a customer connects their on-premises VMware environment to Azure and AVS:

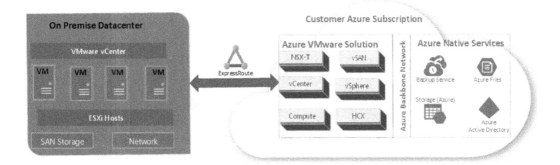

Figure 12.1 – AVS high-level architecture

Since AVS and Azure-native services share the Microsoft Azure backbone, connectivity from AVS to Azure-native services is at the customers' fingertips. This makes it easier for customers to integrate AVS with Azure-native services such as Microsoft Azure Backup, Security Center, Azure Arc, Azure Monitor, and others.

Throughout the rest of this chapter, we will focus on the holistic governance, security, and compliance of your AVS environment. We cannot talk about governance without talking about security and compliance. Governance, security, and compliance are the triad needed for a thriving AVS environment.

We will also look into integrating your AVS environment with your Azure-native services and solutions.

We will look at the components in the following order:

- Security for your AVS environment
- Compliance
- Governance
- Azure-native solutions integration

Let's get started.

Security for your AVS environment

Security is the heart of every solution that Microsoft offers on Microsoft Azure, and AVS is no exception. Suitable security measures must be implemented for your AVS deployments.

The following aspects should be considered when determining which devices, people, or systems may execute tasks within AVS and how to protect the environment holistically.

Security for identity

It is very important to make sure that you integrate your AVS environment with an identity provider. **Active Directory Domain Services** (**AD DS**) or Azure AD DS is utilized by most customers. The following are some key areas that you should focus on:

- **Place limits on permanent access**: In the Azure resource group that hosts the AVS private cloud, AVS uses the Contributor role. To avoid deliberate or accidental contributor rights misuse, limit permanent access. Use a privileged account management system for auditing and determining the duration that highly privileged accounts may be used.

  To manage Azure AD user and service principal accounts, create an Azure AD privileged access group under **Azure Privileged Identity Management** (**PIM**). Create and administer the AVS environment using this group with time-bound, justification-based access.

For administrative tasks, operations, and assignments involving the AVS, use the audit history reports from Azure AD PIM. For long-term audit preservation requirements, the reports can be archived in Azure Storage.

- **Identity management for guest VM**: To enable effective application administration and prevent unwanted access to company data and operations:

  - Connect the AD DS implementation to Azure AD for improved administration and a unified experience for guest authentication and authorization. Make sure to implement your identity management system in a highly available manner to combat any outages of the solution.

  - Log all guest access to all VMs and applications.

- **Centralized identity management**: For management and operation of AVS, utilize domain services-sourced users and groups; do not permit account sharing. It is highly recommended to integrate the VMware vCenter Server and NSX-T Data Center control applications with AD DS or Azure AD DS by using the offered `cloudadmin` account. Create custom vCenter Server roles and link them to AD DS groups to restrict privileged access to VMware private cloud control surfaces at a more granular level.

  Restrict access to the vCenter and NSX-T login pages to only specified subnets in Azure and on-premises. This will prevent rogue actors from attacking those resources.

  The passwords for the `cloudadmin` account on the vCenter Server may be changed or reset using the AVS pane in Azure. You will also need to update the HCX Connector with the new vCenter `cloudadmin` password.

  Every time you employ this break-glass arrangement, rotate these accounts according to your established schedule.

Note

Currently, rotating the password for your NSX-T Manager is not supported by the Azure portal. You will need to submit a support ticket to Microsoft to get this done.

Network security

Security is the center of Microsoft Azure, and it should be the center of your Azure environment. The following are network security recommendations for your AVS environment:

- **Inherent network security features**: Implement network security measures such as traffic filtering, OWASP rule compliance, **distributed denial of service (DDoS)** prevention, and unified firewall administration:

- **Traffic filtering**: As you transition your on-premises security zones to AVS, you should consider implementing network traffic inspection between guest workload segments. You can accomplish this task using either the NSX-T Data Center or a **network virtual appliance (NVA)** of your choice.

- **OWASP Core Rule Set compliance**: Protect the AVS guest web application workloads from generic web assaults. Protect web applications hosted on AVS guest virtual machines using the OWASP capabilities of the Azure **Web Application Firewall (WAF)** or any WAF-capable solution. Enable preventive mode with the most current policy and ensure that WAF logs are integrated into your logging strategy.

- **DDoS protection**: Implementing DDoS protection will safeguard your AVS workloads from assaults that result in monetary loss or a bad user experience. Apply DDoS protection to the virtual network in Azure that hosts the ExpressRoute termination gateway for the AVS connection. Consider implementing DDoS defense automatically using Azure Policy.

- **Unified firewall rule management**: Implement firewall rule management to reduce the risk of unauthorized access caused by duplicate or missing firewall rules. Firewall architecture adds to AVS's more extensive network management and environment security posture. Utilize a managed stateful firewall architecture that permits traffic flow, inspection, centralized rule administration, and event collecting.

- **Ingress internet request logging for guest VMs**: Use an Azure Firewall or an authorized **Network Virtual Appliance (NVA)** that keeps audit logs for inbound requests to guest VMs. Import these logs into your **security incident and event management (SIEM)** system for monitoring and alerting purposes. Utilize Microsoft Sentinel to process Azure event data and logs before integration with current SIEM systems.

- **Session monitoring for egress internet connection security**: To discover unexpected or suspect outbound internet activities, utilize rule control or session auditing of AVS's outgoing internet access. Determine when and where to deploy network inspection for outgoing traffic to achieve maximum security.

Use customized firewall, NVA, and **virtual wide-area network (Virtual WAN)** services for outbound internet access. You can also use the new release feature of Public IP Capability for AVS. The following are some of the features of Public IP Capability in AVS:

- Direct inbound and outbound internet access for AVS to the NSX-T Edge

- The ability to receive up to 1,000 or more public IPs

- DDoS security protection against network traffic in and out of the internet

- Enable support for VMware HCX (migration tool for VMware VMs) over the public internet

- **Controlled access to the vCenter Server**: Unrestricted access to the VMware vCenter Server for the AVS might extend the attack surface. You can securely utilize a dedicated **privileged access workstation** (**PAW**) to access the AVS vCenter Server and NSX-T Manager. Please create a user group and include individual user accounts in it.

- **Secure backups for AVS**: Utilize RBAC and delayed deletion to avoid the purposeful or accidental destruction of backup data required for your AVS data recovery. Utilize Azure Key Vault to manage encryption keys and restrict access to where backup data is stored to reduce the risk of destruction.

Use Azure Backup or another backup system verified for AVS that offers encryption in transport and at rest. Utilize resource locks and soft deletion when utilizing Azure Recovery Services vaults to prevent inadvertent or purposeful backup destruction.

VM and guest application security

Having security in place for your AVS VMs and the applications that run on them is crucial. The following are recommendations on how to secure these resources:

- **Encryption for the guest VMs**: Data-at-rest encryption is provided by AVS for the underlying vSAN storage infrastructure. Data protection measures for some workloads and settings with filesystem access may call for higher levels of encryption. Consider implementing encryption for the guest VM **operating system** (**OS**) and data in certain circumstances. To encrypt guest VMs, you can use a solution such as Azure Disk Encryption for Windows VMs

  The encryption keys should be kept and secured using Azure Key Vault.

- **Database encryption and monitoring**: Encrypt SQL and other databases in AVS to prevent unauthorized access to data in the event of a data breach. Use encryption-at-rest solutions such as **transparent data encryption** (**TDE**) or any other native database capability for database workloads. Verify that workloads use encrypted disks and that those important keys are kept in a key vault belonging to the resource group. Azure Key Vault is recommended for this.

  Reduce the risk of an insider attack by identifying any unexpected database activity. Use native database monitoring, such as Activity Monitor, or an approved partner solution for AVS. Consider utilizing the database capabilities of Azure to improve auditing controls.

- **Advanced threat detection**: Utilize endpoint security protection, security alert configuration, change control mechanisms, and vulnerability assessments to prevent various security threats and data breaches. Microsoft Defender for Cloud is a very good option for threat management, endpoint protection, security alerts, operating system patching, and a consolidated view of regulatory compliance enforcement.

When deploying new guest VMs or before migration, install the Log Analytics agent on VMware vSphere VMs. Set up an Azure Log Analytics workspace for the MMA agent to submit metrics and logs. After the migration, confirm that Azure Monitor and Microsoft Defender for Cloud receive reporting alerts from the AVS VM.

As an alternative, implement a solution from a partner certified for AVS to evaluate VM security postures and ensure regulatory compliance with **Center for Internet Security (CIS)** criteria.

Onboard your guest virtual machines using Azure Arc for servers. Once onboarded, utilize Azure Log Analytics, Azure Monitor, and Microsoft Defender for Cloud to gather logs and metrics and construct dashboards and alerts. Utilize Microsoft Defender Security Center to protect and detect dangers posed by virtual machine guests.

- **Security analytics**: Cyberattacks may be discovered using unified security event collection, correlation, and analytics from the AVS VMs and other sources. Use Microsoft Sentinel as a data source for Microsoft Defender for Cloud. Set up Azure Resource Manager, a **Domain Name System (DNS)**, Microsoft Defender for Storage, and other Azure services necessary for implementing AVS. Consider utilizing a certified partner's data connector solution for AVS.

- **Code security**: This is used to mitigate security vulnerabilities in AVS workloads, including security measures in DevOps workflows. Use current authentication and authorization procedures such as OAuth and OpenID Connect.

  Use *GitHub Enterprise Server on Azure VMware Solution* (`https://learn.microsoft.com/en-us/azure/azure-vmware/configure-github-enterprise-server`) for a versioned repository that ensures the integrity of the code base. Deploy build and run agents either in AVS or in a secure Azure environment.

- **Extended Security Update (ESU) keys**: To push and install security updates on AVS VMs, provide and configure ESU keys. ESU keys for the AVS cluster should be configured using the Volume Activation Management Tool.

It is important to understand the compliance needs and objectives of your customer. These needs and goals will differ for each customer based on where they are located in the world and what sector of business they are operating in.

The following section covers some guidelines to help you comply while operating your AVS environment.

Compliance

The following recommendations should be considered and implemented when preparing for your AVS environment and workload VM compliance:

- **Industry or country-specific regulatory compliance**: Avoid costly legal proceedings and fines by ensuring AVS workload VM compliance with country and industry-specific standards.

To meet regulatory requirements, provide firewall audit reporting for TCP port 443/80 (HTTP/S) endpoints.

- **Compliance policy for corporations**: Workload monitoring is critical for AVS VM's adherence to business policies to prevent violations of company norms and regulations. It is recommended that Azure Arc-enabled servers and Azure Policy, or a comparable third-party solution, be used. Assess and manage AVS workload VMs and apps regularly to ensure regulatory compliance with relevant internal and external laws.

- **Microsoft Defender for Cloud monitoring**: To monitor compliance with security and regulatory standards, use the regulatory compliance view in Defender for Cloud. Set up Defender for Cloud process automation to monitor deviations from the desired compliance posture.

- **Compliance for BC/DR**: Ensure mission-critical apps are accessible during a catastrophe by monitoring BC/DR configuration compliance for AVS workload VMs. Utilize Azure Site Recovery or an AVS-certified BCDR solution, which offers replication provisioning at scale, noncompliance status monitoring, and automated repair.

- **Backup compliance for workload VMs**: Monitor AVS workload VM backup compliance to ensure that the VMs are being backed up on the schedules that have been implemented. Utilize a certified AVS partner solution for tracking and monitoring workload VM backups that give a scalable perspective, drill-down analysis, and an actionable interface.

- **Data retention and residency requirements**: AVS does not allow cluster-stored data retention or extraction. When a cluster is deleted, all active workloads and components are terminated, together with all cluster data and configuration information, including public IP addresses. This data is irretrievable.

- **Data processing**: Read and understand the legal terms when you sign up. Pay attention to the VMware data processing agreement for Microsoft AVS customers transferred for L3 support. If a support issue needs VMware support, Microsoft shares professional service data and associated personal data with VMware. From that point on, Microsoft and VMware act as two independent data processors.

> **Important note**
>
> AVS makes no promises that the service's configuration and metadata are limited to the deployed region. If your needs for data residency demand that all data exist in the deployed region, get help from AVS support.

Implement an AVS environment with strong governance throughout your environment's life cycle. This will enable you to explore proposed design elements while your implementation is underway and help your business achieve regulatory criteria.

Governance

Consider applying the following recommendations when designing your AVS environment and guest VM governance.

AVS environment governance

The following governance recommendations will help you to design and implement a robust AVS environment that will allow for successful deployment and expansion when needed:

- **Governance for host quota**: Inadequate host quotas might result in a week's delay in obtaining additional host capacity for expansion or **disaster recovery (DR)** demands. When requesting the host quota, consider growth and DR requirements, and monitor environment growth and maximums regularly to guarantee sufficient lead time for expansion requests. For example, if a three-node AVS cluster requires an additional three nodes for disaster recovery, request a host quota of six nodes. Also, you should anticipate the growth and rate of migration during your planning stage to understand how many nodes are required for complete workload migration.

- **Financial governance**: Costs should be monitored to ensure proper financial responsibility and budget allocation. Use a cost management solution for cost tracking, cost allocation, budget preparation, alerts, and excellent financial control. Use Azure Cost Management and Billing capabilities to build budgets, generate alerts, assign expenses, and create reports for financial stakeholders for Azure invoiced charges. Cost management should be implemented across your entire Azure landscape.

- **vSAN storage space**: Inadequate vSAN storage space can affect SLA assurances. Review and understand the customer and partner obligations outlined in the SLA for AVS. Assign the proper priority and owners to alerts for the Percentage Datastore Disk Used indicator. 25% of unused storage space is required for vSAN storage.

- **Access to the ESXi hosts**: Unlike a VMware environment on-premises, access to the AVS ESXi hosts is restricted. Third-party software that requires ESXi host access may not operate as intended. Identify any AVS-supported third-party software in the source environment that requires access to the ESXi host. For scenarios requiring ESXi host access, submit a support ticket using the AVS support request procedure in the Azure portal.

- **Density and efficiency of ESXi hosts**: Recognize ESXi host usage for a favorable **return on investment (ROI)**. To get the most return on your AVS expenditure, establish a healthy guest VM density and track total node use against it. When monitoring suggests it, resize the AVS environment and give yourself enough time to add nodes. Use the Placement policies to help with this initiative.

- **Governance failure-to-tolerate (FTT)**: Configure FTT parameters proportional to the cluster size to preserve the SLA for the AVS. Adjust the vSAN storage policy's FTT parameter to the correct value when modifying the cluster size to ensure SLA compliance.

- **Storage policy for VM template**: Too much vSAN storage may be reserved due to a default thick-provisioned storage policy. Make VM templates that don't require space reservations and use a thin-provisioned storage policy. Storage resources are more effective for VMs that don't reserve the entire amount upfront.

- **Azure-native services integration**: The public endpoint for Azure PaaS services can cause network traffic to leave the secure confines of your virtual network. Therefore, it's best to avoid it if possible. Utilize a private endpoint to connect to Azure services such as Azure Blob Storage and Azure SQL Database and keep all traffic inside the confines of the virtual network.

- **Governance for network monitoring**: You must monitor your internal network traffic for malicious or unfamiliar traffic, as well as for compromised networks. Implement solutions such as **vRealize Network Insight** (**vRNI**) and **vRealize Activities** (**vROps**) for comprehensive visibility into the AVS networking operations.

- **Alerts for Service Health, planned maintenance, and security**: Understand and monitor service health to effectively plan and respond to outages and problems. Configure Service Health notifications for AVS service outages, scheduled maintenance, health advisories, and security advisories. Plan and schedule AVS workload operations outside of Microsoft-recommended maintenance schedules.

Governance for workload applications and VMs

Having the correct awareness of your AVS workload security posture VMs enables you to comprehend cybersecurity preparation and reaction and give comprehensive security protection for guest VMs and applications. The following are recommendations for you to implement in your AVS environment:

- **Logging and monitoring of workload VMs**: Before migrating or when adding new workload VMs to the AVS environment, deploy the Log Analytics agent (MMA) on those VMs. Create an Azure Log Analytics workspace and connect it to Azure Automation by configuring the MMA. After migration, use Azure Monitor to confirm the status of any workload VM MMA agents deployed before migration. To troubleshoot OS and application problems more quickly, enable diagnostic metrics and logging on workload VMs. Implement log-gathering and querying features that offer rapid reaction times for troubleshooting and debugging. Enable near-real-time VM analytics on workload VMs to identify operational problems and performance bottlenecks quickly. Set up log alerts to record boundary situations for workload virtual machines.

- **Governance for adding a workload VM to a domain**: Use extensions such as `JsonADDomainExtension` or similar automation options to minimize error-prone manual operations when enabling AVS guest VMs to join an Active Directory domain automatically.

- **Update/patching governance for workload VMs**: Implement a robust patch management policy for all your AVS workload VMs. They are leading attack vectors that might expose or compromise your AVS workload VMs and applications resulting from delayed or insufficient upgrades or patching. Make sure that guest VM updates are installed on time.

- **Backup governance for workload VMs**: Schedule frequent backups to avoid missing or relying on outdated backups, which can result in data loss. Utilize a backup system that can automate backups and track their success. Monitor and generate alerts for backup-related events to verify that planned backups succeed. Use Microsoft Azure Backup Server or any other AVS-certified backup solution.

- **BCDR governance for workload VMs**: During **business continuity and disaster recovery (BCDR)** events, undocumented **recovery point objective (RPO)** and **recovery time objective (RTO)** requirements can result in bad customer experiences and missed operational goals. Implement DR orchestration to avoid business continuity delays.

- **Enable Microsoft Defender for Cloud**: For running Azure services and AVS application VM workloads.

- **Use Azure Arc-enabled servers**: For managing AVS guest VMs with tools that replicate Azure-native resource tooling, including the following:

  - Azure Policy to govern, report, and audit guest configurations and settings

  - Azure Automation State Configuration and supported extensions to simplify deployments

  - Update Management to manage updates for the AVS application VM landscape

  - Tags to manage and organize the AVS application's VM inventory

Utilize a DR solution for your AVS environment that offers DR orchestration, identifies and reports any faults or failures with continuous replication to a DR site, and provides DR orchestration. VMWare SRM, Zerto, and JetStream are some BCDR options for AVS. Document RPO and RTO requirements for Azure and AVS applications. It is highly recommended that you implement a secure Azure VMware Solution with robust governance in your environment throughout its life cycle. This will assist your company in meeting regulatory standards and allowing you to explore suggested design components while your implementation is in progress.

Azure-native solutions integration

One of the key features of AVS is the capability to integrate the solution with Azure-native solutions. The connectivity from AVS to the Azure-native environment is done through the included 10 GiB ExpressRoute circuit. Having both environments sit in a Microsoft data center allows connectivity while utilizing the Microsoft Azure backbone. AVS can integrate with Azure-native services without leaving Azure. This is important to understand as the latency is in the sub-millisecond range, making for an excellent user experience.

In this section, we will walk through the process of attaching a VM in your AVS environment to an Azure file share.

Azure Files provides fully managed file shares in the cloud, accessible through the **Server Message Block** (**SMB**) protocol, the **Network File System** (**NFS**) protocol, and the Azure Files REST API. Cloud-based and on-premises deployments can simultaneously mount Azure file shares. SMB Azure file shares are available to Windows, Linux, and macOS clients. Linux and macOS clients can gain access to Azure NFS file shares. In addition, SMB Azure file shares may be cached on Windows servers using Azure File Sync to access where the data is consumed quickly.

Use cases for Azure Files

The following are some use cases for Azure Files:

- **Application migration**: Applications that rely on file shares to store user or application data may be easily migrated to the cloud using Azure Files. Both the "traditional" migration scenario, in which the application's data is moved to Azure, and the "hybrid" migration scenario, in which the application's data is moved to Azure Files, but the application remains on-premises, are enabled by Azure Files.

- **Replace or supplement traditional on-premises file servers**: Azure Files can be used to replace or augment traditional on-premises file servers or **network-attached storage** (**NAS**) devices. Azure File Sync may replicate SMB Azure file shares to Windows servers on-premises or in the cloud for performance and distributed caching of the data. Popular OSs, including Windows, macOS, and Linux, may immediately mount Azure file shares from anywhere globally. Azure Files AD Authentication enables AD DS deployed on-premises to provide access control for SMB Azure file sharing.

- **Reduce the TCO of AVS**: Because Azure Files can be used to replace or augment the traditional file servers, instead of migrating your on-premises files servers to AVS, you can instead utilize Azure Files. This will free up your expensive vSAN for more IOPs-driven workloads.

Key advantages

Some of the key advantages of utilizing Azure Files are as follows:

- **Fully managed solution**: Creating Azure file sharing does not need hardware or operating system management. Azure Files eliminates needing to patch the server's operating system with critical security updates or replace defective hard drives.

- **Access sharing**: The industry-standard SMB and NFS protocols are supported by Azure file shares, so you can easily replace your on-premises file shares with Azure file shares without worrying about application compatibility. Applications that require shareability benefit significantly from the ability to share a filesystem across several computers, programs, and application instances.

- **Availability**: Azure Files was developed from the ground up to be always accessible. With Azure Files in place of on-premises file sharing, network disruptions and local power outages are no longer a concern.

- **Ease of use**: When an Azure file share is mounted on your PC, accessing the data is as simple as going to the location where the file share is mounted and opening or editing a file.

- **Multiple toolsets**: As part of the management of Azure applications, Azure file shares may be created, mounted, and managed using PowerShell cmdlets and the Azure CLI. You can create and manage Azure file shares using the Azure portal and Azure Storage Explorer.

- **Familiar programmability**: Applications running in Azure can access data in the share via filesystem I/O APIs. Developers can therefore leverage their existing code and skills to migrate existing applications. In addition to System I/O APIs, you can use Azure Storage Client Libraries or the Azure Files REST API.

How to create an Azure file share

Before creating an Azure file share, you need to make sure that you think about how you will use this solution. Answer the following questions before creating the new Azure file share:

- **What is the size of the file share?** Azure file shares in local and zone redundant storage accounts may grow up to 100 TB. In geo and geo-zone redundant storage accounts, Azure file shares are limited to 5 TiB in size.

- **What are the performance requirements for the file share?** There are two types of Azure file shares: Standard and Premium. The Standard file share is deployed to **hard disk drive (HDD)** hardware, while the Premium file share is deployed to a **solid-state drive (SSD)**.

   For high IOPs workloads, deploy the premium Azure files.

- **What are the redundancy requirements for the file share?** While geo-redundant, geo-zone-redundant, locally redundant, or zone redundant storage are all options for standard file shares, only locally redundant and zone redundant file shares enable the large file share functionality. Geo-redundancy is not supported in any way by premium file sharing.

   In a small number of Azure regions, premium file sharing is offered with locally and zonally redundant services.

Prerequisites

The following are the prerequisites to creating an Azure file share:

- A storage account created with the premium option selected for *Performance*
- The same storage account created with the File shares option selected for *Premium account type*
- TCP port 445 open on your virtual network (this is only for the SMB file share type)

The following screenshot shows how to create a storage account with the different disk performance types:

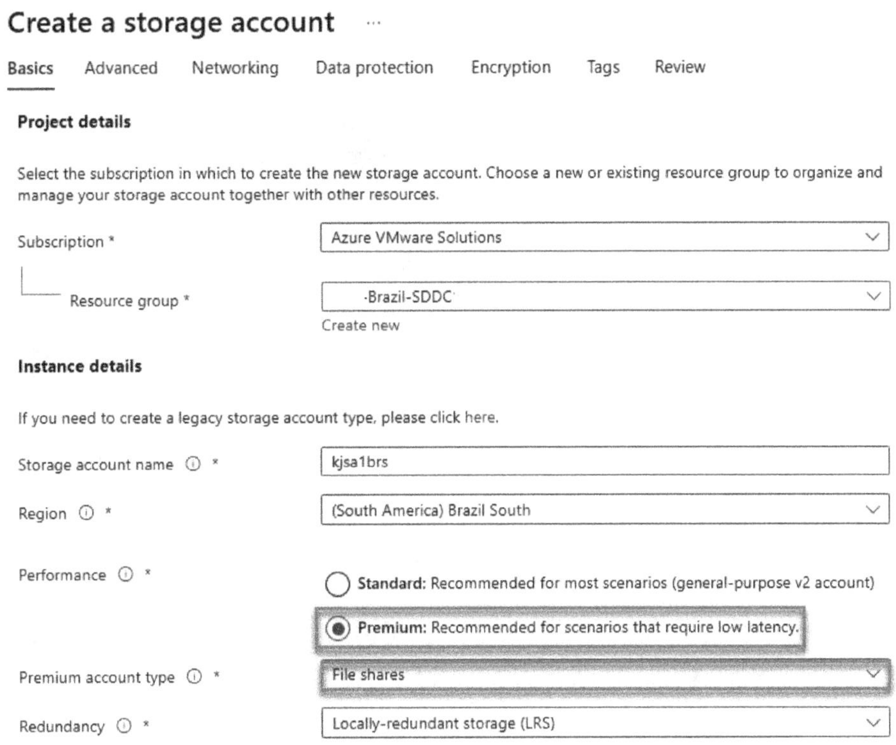

Create a storage account ...

Basics   Advanced   Networking   Data protection   Encryption   Tags   Review

Project details

Select the subscription in which to create the new storage account. Choose a new or existing resource group to organize and manage your storage account together with other resources.

Subscription *

> Azure VMware Solutions

Resource group *

> ·Brazil-SDDC·
>
> Create new

Instance details

If you need to create a legacy storage account type, please click here.

Storage account name ⓘ *

> kjsa1brs

Region ⓘ *

> (South America) Brazil South

Performance ⓘ *

> ○ Standard: Recommended for most scenarios (general-purpose v2 account)
>
> ◉ Premium: Recommended for scenarios that require low latency.

Premium account type ⓘ *

> File shares

Redundancy ⓘ *

> Locally-redundant storage (LRS)

Figure 12.2 – Storage account options

Creating an Azure file share

After creating your storage account, you can create your file share. This procedure is essentially the same, whether you're creating a Premium or a Standard file share. The following differences should be considered when creating a file share.

Standard file shares can be installed in one of the standard tiers: transaction optimized (the default), hot, or cool. This is a per-file-share tier that is unaffected by the blob access tier of the storage account (this property only relates to Azure Blob storage – it does not relate to Azure Files at all). After the share has been deployed, its tier can be changed at any moment. Premium file shares cannot be switched to a standard tier directly.

Follow these steps to create an Azure file share in an existing storage account:

1. Log into your Azure portal and search for the storage account where you will create the file share.

2. From the storage account under **Data storage**, select **File shares** | **+ File share**:

Figure 12.3 – Adding a file share

3. In the **New file share** window, enter a name for the file share, and enter the size of the file share for **Provisioned capacity** (I used 1,024 GiB). Specify the storage protocol (SMB or NFS). Click on **Create**:

Figure 12.4 – Creating a new file share

You will see the new file share that was created. Take note of the **Protocol** and **Provisioned capacity** details:

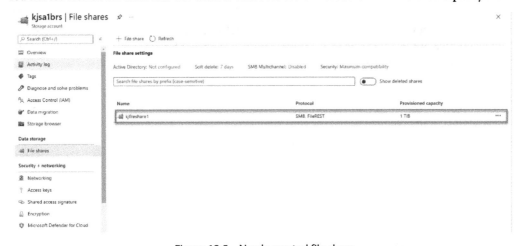

Figure 12.5 – Newly created file share

Mapping the Azure file share to an AVS VM

Now, we will map the newly created Azure file share to a virtual machine in your AVS environment:

1. In the Azure portal, browse to the Azure storage account where you created the Azure file share.

2. Click on **File shares,** click on the three dots (...) on the file share that was created earlier, and select **Connect**:

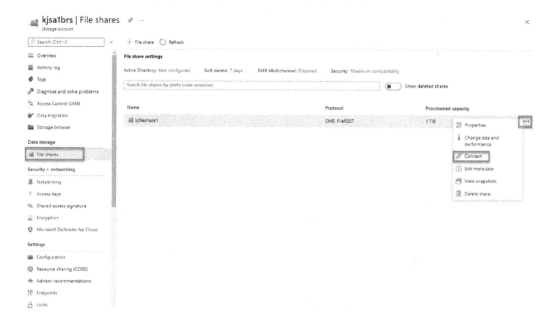

Figure 12.6 – Connecting a file share to a VM

3. In the **Connect** window, make sure **Windows** is selected as the operating system type. Choose a **Drive letter** and select a **Storage account key**. Then, click on **Show Script**. Copy the displayed script and save it to Notepad on your local computer:

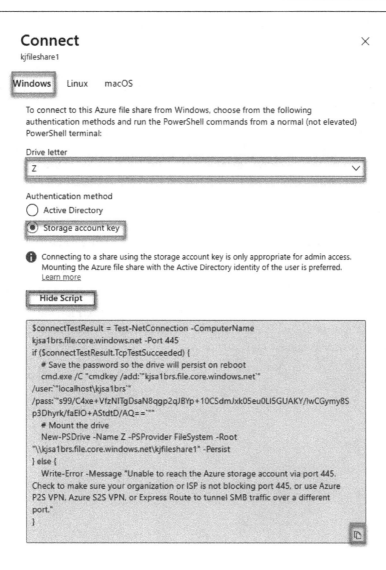

Figure 12.7 – Connecting a file share to a VM

4. Connect to the virtual machine that is hosted in your AVS environment. Paste the copied script into PowerShell and press *Enter* to run the command. You will see that you have a mapped network drive with the drive letter you chose.

Azure file share is just one of the many Azure-native solutions you can integrate with your AVS environment. With the proximity of Azure native and AVS, customers are starting their application modernization the moment they migrate to AVS.

Summary

Throughout this chapter, the focus was on how to implement a unified security and compliance approach for your AVS environment. All your AVS infrastructure can use the same set of apps, allowing for streamlined management, better security, and a more consistent user experience.

Take advantage of Azure's scalability, performance, and innovation while using VMware technologies such as vCenter Server, vSAN, and NSX-T Manager. AVS's security and privacy can be improved by using vSphere's role-based access control. Integration between vSphere SSO LDAP functionalities and Azure Active Directory is possible.

Monitor for vulnerabilities, transform risk vectors into regulating corporate policies, and expand governance rules so that they include Cost Management, Security Baseline, Identity Baseline, Resource Consistency, and Deployment Acceleration to evaluate and manage risk tolerance.

Integrate your AVS environment with Azure-native solutions that will help you to secure your private cloud and solutions that will help you to offset the cost.

The next chapter is the last chapter of this book. There, we will review the critical themes covered in the preceding chapters. In addition, we will examine the AVS roadmap and best practices for developing, deploying, and maintaining your AVS infrastructure.

Summary of Azure VMware Solution, Roadmap, and Best Practices

AVS is a first-party Microsoft Azure service developed in conjunction with VMware that provides a familiar vSphere-based, single-tenant private cloud on Azure that is like the one used by VMware. The VMware technology stack consists of the following components: vSphere, NSX-T, vSAN, and HCX. AVS is installed on dedicated infrastructure in Azure data centers and runs natively on that infrastructure. In comparison with existing on-premises VMware infrastructures, AVS provides a consistent and well-known user experience. Customers may deploy an AVS environment in a matter of hours and migrate VM resources in a matter of minutes. Microsoft supplies all the networking, storage, management, and support services that are required.

The following is a high-level architectural overview of how a customer connects their on-premises VMware environment to Azure and AVS. ExpressRoute Global Reach, a capability that connects numerous ExpressRoute circuits, is used to connect your on-premises environment to your AVS private cloud.

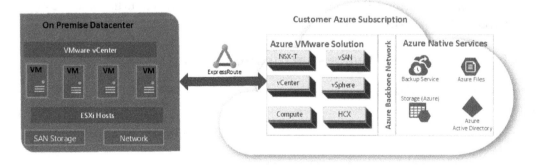

Figure 13.1 – Relationship between private clouds and VNets

Throughout this book, we have looked at what AVS is, the use cases for AVS, the networking design, as well as management and security for AVS. This chapter will bring everything together. We will also be outlining some kind of roadmap for AVS and talking about some of the best practices based on customers' experiences of deploying and managing AVS.

We will be looking at the following topics in this final chapter:

- AVS overview
- Use cases for AVS in an enterprise
- Network and connectivity topology for AVS
- AVS roadmap
- Best practices for planning, deploying, and managing AVS

AVS overview

AVS provides a private cloud environment that can be accessed from both on-premises and Azure-based infrastructure resources. Azure ExpressRoute, VPN connections, or Azure Virtual **Wide-Area Network** (**WAN**) are options for connectivity. However, to make these services available, specific network address ranges and firewall ports must be configured.

When a private cloud is deployed, private networks are formed for management, provisioning, and vMotion. These private networks will be used to connect to vCenter and NSX-T Manager, as well as to perform VM vMotion and deployment. The private network must be a /22 CIDR. This /22 is only used for the management components and not for your workload segments. You will need additional networks for your workloads.

It is possible to link private clouds to on-premises systems using ExpressRoute Global Reach. It establishes direct connections between circuits at the **Microsoft Enterprise Edge** (**MSEE**). Your subscription must have a **Virtual Network** (**vNet**) with an ExpressRoute circuit to on-premises for the connection to work. The reason for this is that vNet gateways (ExpressRoute gateways) are unable to transfer traffic across circuits. This implies that you can connect two circuits to the same gateway, but the traffic will not be transferred from one circuit to another.

AVS hardware and software specification

AVS private clouds and clusters are constructed on top of dedicated bare-metal hyper-converged Azure infrastructure hosts.

The hardware and software configuration for the different host types that are currently available are as shown:

SKU	CPU (GHz)	RAM (GB)	vSAN Cache Tier (TB, raw)	vSAN Capacity Tier (TB, raw)	Network Interface Cards (NICs)	Regional availability
AV36	Dual Intel Xeon Gold 6140 CPUs with 18 cores/CPUs @ 2.3 GHz, total of 36 physical cores (72 logical cores with hyperthreading)	576	3.2 (NVMe)	15.20 (SSD)	4x 25 Gb/s NICs (2 for management and control plane, 2 for customer traffic	All product regions
AV36P	Dual Intel Xeon Gold 6240 CPUs with 18 cores/CPUs @ 2.6 GHz / 3.9 GHz Turbo, total of 36 physical cores (72 logical cores with hyperthreading)	768	1.5 (Intel Optane Cache)	19.20 (NVMe)	4x 25 Gb/s NICs (2 for management and control plane, 2 for customer traffic)	Selected regions (*)
AV52	Dual Intel Xeon Platinum 8270 CPUs with 26 cores/CPUs @ 2.7 GHz / 4.0 GHz Turbo, total of 52 physical cores (104 logical cores with hyperthreading	1536	1.5 (Intel Optane Cache)	38.40 (NVMe)	4x 25 Gb/s NICs (2 for management and control plane, 2 for customer traffic)	Selected regions (*)

Table 13.1 – AVS nodes hardware SKUs

The Azure pricing calculator will detail the regional availability of these new AVS nodes.

The AV36P and AV52 are new additions to the AVS node SKU. The AV36 was the only available SKU until mid-2022. With the introduction of the new AVS nodes, customers will have more options to choose from based on their workload specifications that will be migrated to AVS.

A minimum of three hosts are required for an AVS cluster. Only hosts of the same SKU may be used in a single AVS private cloud. Multiple clusters with various host types may be used in a single AVS environment. Hosts needing to establish or scale clusters are drawn from a separate pool of hosts. Before joining a cluster, the hosts will need to pass hardware checks. In addition, all data will be wiped.

You may use the Azure portal or the Azure CLI to create new private clouds or scale existing ones.

Microsoft also upgraded the software version of VMware on AVS in mid-2022. The following table shows the latest software specifications:

AVS Software Specification	
ESXi	7.0 U3c
VMware vCenter Server	7.0 U3c
vSAN	7.0 U3c
vSAN on-disk format	10
VMware NSX-T Data Center	3.1.2 Advance
HCX	4.4.2

Table 13.2 – AVS software specification

AVS Enterprise is now available to customers without any additional cost.

We will now take a look at the architecture of AVS at a high level.

AVS high-level architecture

Each AVS environment is deployed with its own 10 GB ExpressRoute circuit (and thus its own virtual MSEE device), which allows you to connect the customers' ExpressRoute circuit using Global Reach. The normal use of Global Reach incurs a cost. However, when used for AVS and configured from the Azure portal, the cost is reduced.

See the following high-level AVS networking overview where a customer's on-premises environment is connected to Azure using an ExpressRoute circuit, which is terminated in an Azure VNet on an ExpressRoute Gateway. The ExpressRoute circuit is also connected to the AVS ExpressRoute circuit utilizing Azure Global Reach. There is also a connection from the ExpressRoute gateway to the AVS ExpressRoute circuit. See the following high-level AVS networking overview:

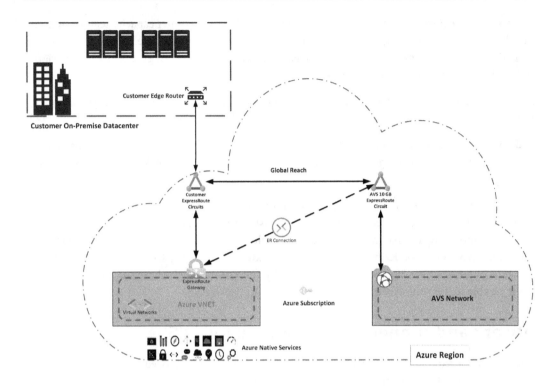

Figure 13.2 – Overview of high-level AVS networking

The architecture in *Figure 13.2* is the most commonly used architecture for connecting from AVS to Azure and also to the customer's on-premises environment.

There are multiple use cases for a customer to migrate their workloads to AVS. In the next section, we will look at some of those use cases.

Use cases for AVS in an enterprise

You can migrate your VMware workloads from your on-premises data center to AVS and integrate additional Azure services with ease using the same VMware tools that you are already familiar with. However, while there are other advantages, we've identified the top five reasons why AVS is proving to be the most cost-effective path to the cloud for many enterprises:

Data center footprint deduction, consolidation, and retirement: Nowadays, we see many customers reducing their on-premises data center footprint for many reasons, including cost, getting out of managing data centers, and focusing more on their business. AVS helps customers reduce the size of their data center's footprint by redeploying their VMware-based VMs on a one-time basis.

Additional data center: Customers are now able to increase their data center capacity seamlessly and elastically—while also adjusting their cost-on-demand for short periods of time. We see this kind of need in the logistic business where customers need to increase their data center capacity for a period and then decrease that capacity once the need is no longer required.

Disaster recovery and business continuity: AVS can be used as a primary or secondary data center without the management overhead.

Speed and simplification of migration/hybrid cloud: AVS has proven to be one of the most efficient and straightforward methods of getting started on Azure without having to make any changes to your existing apps or servers.

AVS is very cost-effective: When it comes to running VMware apps on Windows Server and SQL Server, AVS is the most cost-effective option. If you use your on-premises data center effectively, you can save money by not having to purchase multiple licenses for both on-premises and cloud applications. WS 2008/2008 R2 has slightly different **Extended Security Updates** (**ESU**) duration coverage as opposed to WS 2012/2012 R2. Additionally, for customers who have software assurance, you will be able to use your existing Windows and SQL license in AVS.

Network and connectivity topology for AVS

Although it is not required, it is highly recommended to have already deployed an Azure enterprise-scale landing zone before implementing an AVS private cloud. This will provide a better management experience for future deployments that accounts for scale, security governance, networking, and identity. It is best to have that in place before deploying AVS, as this will help you connect your AVS infrastructure to your on-premises data center, AVS to Azure, and AVS to the internet.

Implementing a VMware **software-defined data center** (**SDDC**) with the Azure cloud ecosystem has some unique design challenges to think about when planning for your deployment.

Some of the challenges include the following:

- Hybrid connectivity to facilitate the connection between on-premises data center, Azure, edge networks, and global users

- Reliability and performance to scale workloads and maintain low latency and a consistent experience

- A zero-trust network security model for a cloud environment that entails all the network security that is needed

- Extensibility for ease of expanding your networks without the need for re-architecture

We will now look at the different networking components and concepts used to create the different connectivity medians for AVS:

- **Azure Vnet**: Your Azure private networks are built on VNet. When a virtual network is set up, it behaves and seems much like a traditional network in your own data center. However, it has the scalability, availability, and isolation benefits of the Azure architecture. When using Azure Vnet, various Azure resource types, including VMs and databases, can connect safely and securely to one another, the internet, and on-premises data centers.

- **Hub-spoke network topology**: In this topology, the virtual hub network serves as the central connection point for multiple spoke virtual networks. A spoke virtual network that connects to the hub can be used to separate different types of workloads from each other. An on-premises data center, AVS SDDC, can also be linked up to a hub through a connection point (ExpressRoute and or a **site-to-site (S2S)** VPN).

- **Network Virtual Appliance (NVA)**: This is a virtual appliance that provides WAN optimization, security, connectivity to different endpoints, application delivery, and so on. Some examples of an NVA include F5-BigIP, Azure Firewall, Cisco Firewall, and Barracuda Firewall. An NVA in Azure functions the same way a physical appliance does in a customer data center.

- **Azure Virtual WAN (vWAN)**: vWAN is a unique networking service that you can use to integrate many features such as networking, routing, and security functionalities to provide a single interface for operation.

  Some of the functionalities of Azure vWAN include S2S VPN connectivity, ExpressRoute connectivity, which is a private connection, routing, and Azure Firewall. It also includes encryption for private connectivity. You can start with just one use case, and then add functionalities as they are needed.

  The architecture for Azure vWAN is a hub and spoke architecture that can scale as needed by adding additional spokes.

- **Layer 4 (L4)**: The fourth layer of the OSI model is referred to as L4. It is also known as the **Transport Layer**. L4 enables data to be transmitted or transferred between hosts or end systems transparently. Error recovery and flow control are both handled by L4. The following are some of the protocols used in L4:

  - **Transmission Control Protocol (TCP)**
  - **Multipath TCP (MPTCP)**
  - **User Datagram Protocol (UDP)**
  - UDP-Lite
  - **Reliable UDP (RUDP)**
  - **AppleTalk Transaction Protocol (ATP)**
  - **Sequenced packet exchange (SPX)**

- **Layer 7 (L7):** The application layer, L7, is the final layer of the OSI model and is the highest layer. Layer 7 identifies the communication parties and the level of service between them. It is L7's job to keep data private and authenticate users, and it does so by looking for any limits on the data syntax. This layer is responsible for all API interactions. The following are some of the main protocols of L7:

  - HTTP

  - HTTPS

  - SMTP

Understanding networking requirements for AVS

Setting up the landing zone for AVS requires a thorough understanding of Azure network design and implementation techniques. A wide range of capabilities is supported by Azure networking products and services. How to arrange services and choose the right architecture relies on your organization's workloads, governance, and requirements since every organization is different.

In the following, you will find some essential requirements and considerations that will affect your AVS deployment decisions:

- Connectivity from on-premises data centers to AVS – will you be connecting over ExpressRoute or an S2S VPN, and will ExpressRoute Global Reach be enabled?

- Will AVS be connecting to an Azure vNet hub for connectivity to Azure native services or to a vWAN hub?

- L2 extension from the on-premises data center to AVS (this is done to retain VM IP addresses).

- Do you have an NVA in your current Azure environment?

- Will applications require HTTP/S or not for internet ingress?

- Traffic inspection needs the following:

  - AVS access to Azure native services

  - AVS access back to the on-premises data center

  - Egress access to the internet from AVS

  - Ingress access from the internet to AVS

  - Traffic connection with AVS

Networking scenarios for AVS with traffic inspection

Default route propagation enabled in a secured virtual WAN hub will benefit customers from this design if the following applies:

- Between AVS and their on-site data center, there is no need for traffic inspection.
- There is no need for traffic inspection between AVS and their Azure vNet.
- Examining the traffic between AVS and the internet is required.
- In this design, the customer will have to add services for L4 and L7 ingress if they so require. We are also assuming that the customer already has an ExpressRoute connection in place between their on-premises data center and Azure.

You can implement this architecture with the following components:

- Application Gateway for L7 load balancing and SSL offloading
- An Azure Firewall in the secure vWAN hub (or any other NVA)
- Configure L4 **destination network address translation** (**DNAT**) on Azure Firewall to filter and translate ingress network traffic
- Configure all egress traffic through Azure Firewall on the vWAN hub
- Implement ExpressRoute, SD-WAN, or a VPN connection between AVS and the on-premises data center

The following diagram illustrates *scenario 1*:

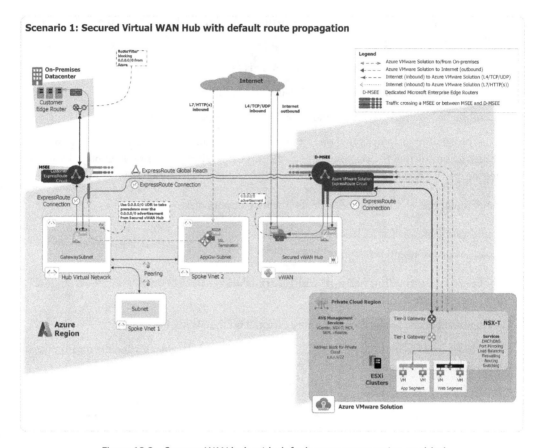

Figure 13.3 – Secure vWAN hub with default route propagation enabled

Things to consider

If the default 0.0.0.0/0 route that is being advertised from AVS is interfering with your existing environment, you will need additional steps to prevent route propagation.

To solve this problem, you can do the following:

- Use an on-premises edge device to block the 0.0.0.0/0 route
- Otherwise, try the following:
  - Disconnect the ExpressRoute, VPN, or virtual network from the secured vWAN hub
  - Reconnect those connections once the 0.0.0.0/0 route is disabled

Enterprise-scale for AVS is an open source set of Azure Resource Manager and Bicep templates for planning and deploying AVS. You may consider it a template for building a scalable AVS that can scale up in the future. This open source solution explains how to construct a scalable AVS environment using Azure landing zone subscriptions. It also uses an example to demonstrate how to set up subscriptions. With a focus on large-scale deployment design concepts, the implementation follows the architecture and best practices of the Cloud Adoption Framework's Azure landing zones.

Planning for an AVS Deployment

Building VMs and migrating successfully requires a production-ready environment, which can only be achieved with careful planning of your AVS deployment. You'll identify and gather the many pieces of information for your deployment as you go through the planning phase. Please keep a record of the data you gather as you plan so you can refer to it while deploying. You'll have a production-ready environment for VM creation and migration following a successful deployment.

See the components below for a successful AVS deployment:

- Select the region, Azure subscription, resource group, and name of the resource for your AVS environment.

- Depending on the size of the hosts, determine the number of clusters and hosts needed for your deployment.

- For an eligible Azure plan, request a host quota (eligible Azure plans are EA, CSP, MACC, and MCA).

- Request a /22 address space from your networking team. This will be used for the management components of AVS.

- Request a single network from your networking team for your workload segment. This can be a /24 address space.

- You will need an ExpressRoute gateway if one does not already exist in your VNet hub in Azure.

Subscription identification

One thing you will need to do is to identify the subscription that you will be deploying the AVS in. An Azure subscription is a logical grouping of Azure services associated with an Azure account. You will need to have a subscription to use Azure's cloud-based services since it acts as a single billing unit for Azure resources used in that account.

An Azure subscription is linked to a single account used to create the subscription and is used for billing purposes. A subscription can contain numerous resources.

You can have many subscriptions for various reasons, including billing, because each subscription creates its own set of billing reports and invoices.

You can use an existing subscription or create a new one for AVS.

The following is an illustration of the relationship of an Azure subscription with other Azure components:

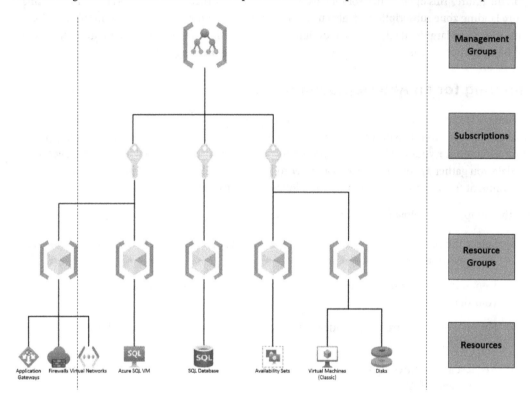

Figure 13.4 – Azure subscription flow

Resource group identification

After the subscription is identified, you will now need to decide on the resource group. You can either utilize an existing resource group or create a new one specifically for AVS.

A resource group is an Azure solution container that stores related resources. The resource group can contain all the solution's resources or just the ones you want to manage as a group. Based on what makes the most sense for your company, you select how to allocate resources to resource groupings. Add resources with the same lifetime to the same resource group to make it easier to publish, update, and delete them.

The metadata about the resources is stored in the resource group. As a result, when you specify a location for the resource group, you're also specifying the location of the metadata. You may need to verify that your data is stored in a specific location for compliance reasons.

Azure region

An Azure region consists of multiple data centers to provide redundancy and availability of your applications. You create Azure resources in defined geographic regions such as East US, North Central US, or West US. This method allows you to be more flexible when designing apps, allowing you to create solutions that are the most useful to your customers while also meeting any legal, compliance, or tax requirements.

It is possible to have multiple resources communicating with each other in different regions. However, it is highly recommended that all resources for your AVS be deployed in the same region.

Region pairs in Azure

Within the same geography, each Azure region is associated with another. This strategy provides for resource replication across geographies, such as VM storage, which should lessen the likelihood of natural disasters, civil unrest, power failures, or physical network outages hitting both regions simultaneously. Region pairs also have the following advantages:

- In the event of a more significant Azure outage, one region from each pair is prioritized to help speed up application recovery
- To minimize downtime and the possibility of an application outage, planned Azure updates are rolled out one by one to paired regions

> **Note**
> You can see the full list of Azure regional pairs at `https://docs.microsoft.com/en-us/azure/virtual-machines/regions`.

AVS resource name

The resource name, for example, `ABCPrivateCloud1`, is a polite and descriptive name for your AVS private cloud.

It's critical to note that the name can't be more than 40 characters. You won't be able to create public IP addresses for usage with the private cloud if the name exceeds this limit.

Determining the number of nodes

You'll need to specify the number of hosts you wish to deploy when deploying your AVS environment. Clusters can be added, removed, and scaled. By default, one vSAN cluster is established for each private cloud. Three nodes are the minimum for an AVS cluster.

It is highly recommended that an assessment be done in your on-premises environment to determine the VM count, CPU usage of each VM, and storage usage. There are different tools that you can use for this assessment. Azure Migrate, Movere, and RVTools are examples of assessment tools that you can use.

Once the assessment is done, work with your Microsoft account team where they will do a node count exercise and calculate pricing.

Most other cluster configuration and operation aspects are handled by vSphere and NSX-T Manager. vSAN oversees all local storage on each host in a cluster.

Host quota request for AVS

AVS is not enabled in your Azure subscription by default. Because of this, you will need to submit a support ticket to have your AVS hosts allocated to your subscription for either a new deployment or an existing one.

Requesting a /22 address space for AVS management components

A /22 address space is required for AVS deployments. This address space is then broken up into smaller segments that are used for the management components of AVS. These components include vCenter, NSX-T, HCX, vMotion, and the Tier-0 gateway. The following diagram shows the IP address segments for AVS management:

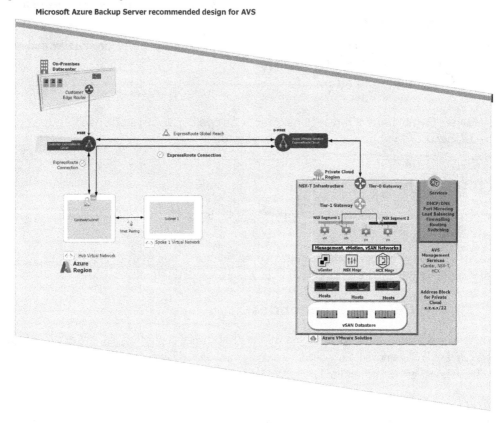

Figure 13.5 – AVS Management segments

Please note that no current network segment on-premises or Azure should overlap with the /22 CIDR network.

Defining the AVS workload network segments

The VMs must connect to a network segment, just as with any other VMware vSphere environment. As the AVS's production deployment grows, it's common to see a mix of on-premises L2 extended segments and local NSX-T network segments. The L2 network is normally extended when customers want to retain their current IP addresses.

In *Figure 13.6*, you can see two NSX segments where the customer workload VM resides in AVS:

Microsoft Azure Backup Server recommended design for AVS

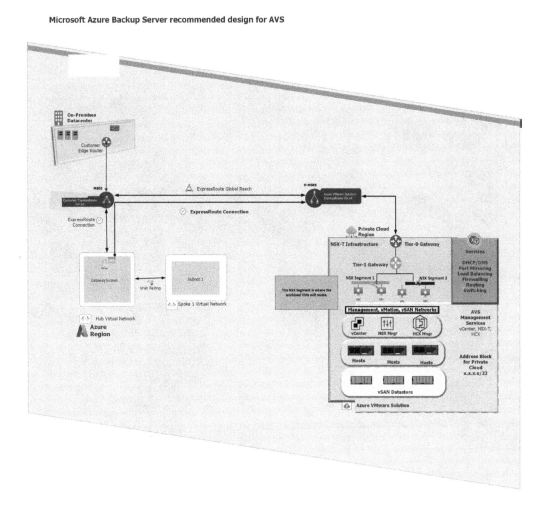

Figure 13.6 – AVS workload segments

Determine a single network segment (IP network) for the initial deployment, such as `10.0.2.0/24`. During the first deployment, this network section is mainly utilized for testing. The address block must not overlap with any network segments on-premises or in Azure, and it must not be within the already specified `/22` network segment.

Defining the virtual network gateway

AVS can be connected to an Azure S2S VPN connection. However, because of the low latency requirements, having a dedicated connection with minimal latency, such as an ExpressRoute circuit, is strongly recommended. The following diagram illustrates an ExpressRoute Gateway connection:

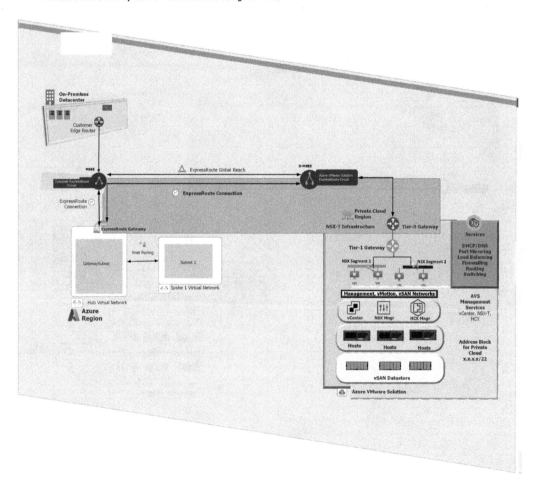

Figure 13.7 – ExpressRoute Gateway connection

To connect AVS to your Azure VNet, you will need to make a connection from the AVS ExpressRoute circuit, you'll also need an Azure Vnet gateway. You can use an existing ExpressRoute Vnet gateway or create a new one if none exists.

This connection gives a customer access to the other Azure services from AVS. You can choose between four different ExpressRoute Gateway SKUs, such as Standard, HighPerformance, UltraHighPerformance, and ErGw1Az – ErGw3Az (a zone-redundant gateway option). These gateways have throughput speeds of 1 Gbps, 2 Gbps, and 10 Gbps, respectively.

Managing an AVS environment

A VMware-verified solution, AVS is tested and certified regularly to ensure compatibility with the latest version of vSphere. Private cloud infrastructure and software are managed and supported by Microsoft for a customer. Microsoft handles everything for you so that you can focus on developing and deploying workloads using the native Azure resources and your private cloud. You can rest assured that the continuous updates for AVS private cloud and VMware software guarantee that your deployed private cloud is up to date with the most recent security, stability, and feature sets.

Microsoft frequently updates the shared accountability diagram for IaaS, PaaS, and SaaS-based services. AVS additionally includes a responsibility allocation matrix. As you can see, Microsoft abstracts a large chunk of continuous maintenance, security, and administration, leaving your business free to focus on more strategic goals, including the deployment of guest OSes and VMs. Life cycle process and configuration management solutions that can be deployed on Azure should also be taken into account. With this setup, Microsoft takes on more of the burden of maintaining the underlying infrastructure for AVS, relieving you of some of the burdens typically associated with doing so.

Microsoft handles the underlying infrastructure when AVS is implemented in Azure. This responsibility modifies standard operating procedures and process flows for IT departments where central IT staff manages the on-premises VMware infrastructure and will need to adjust their standard operating procedures and process flow as a result of this duty. The IT department is unable to do things such as acquire root access to individual ESXi servers. Because of this strategy, operations engineers can stop worrying about the mundane tasks of maintaining a standard VMware environment and instead concentrate on developing new features for the applications and workloads that are the foundation of their company's mission. The scope of the digital transition is expanded much beyond that of audiovisual systems.

The following table shows the shared responsibilities of both Microsoft and customers.

	Deployment	Life cycle	Configuration
Physical infrastructure	Microsoft	Microsoft	Microsoft
Physical security	Microsoft	Microsoft	Microsoft
Azure/AVS portal	Microsoft	Microsoft	Microsoft
Hardware failure	Microsoft	Microsoft	Microsoft
ESXi host	Microsoft	Microsoft	Microsoft
Host patching	Microsoft	Microsoft	Microsoft
NSX-T	Microsoft	Microsoft	Customer
Identity management	Microsoft	Microsoft	Customer
vCenter	Microsoft	Microsoft	Customer
vSAN	Microsoft	Microsoft	Customer
Virtual machines	Customer	Customer	Customer
Guest OS	Customer	Customer	Customer
Applications	Customer	Customer	Customer

Legend: Microsoft / Customer

Table 13.4 – AVS shared responsibility

Leveraging governance for AVS

AVS is a VMware-powered Azure first-party offering that offers vSphere clusters in a single-tenant private cloud environment. The VMware technology stack utilizes Azure's highly secure computing, storage, and networking technologies. Users and applications can access it using both on-premises vSphere solutions and Azure environments or resources.

Connecting to Azure cloud services using a dedicated private and redundant Layer 3 network fiber connection with up to 100 Gbps throughput does not require an ExpressRoute circuit, but it is strongly recommended. You can link your AVS environment to your Azure native environment to access the other Azure native services and solutions.

All private clouds contain vCenter Server, ESXi, vSAN, and NSX-T Data Center, allowing you to move workloads from on-premises vSphere infrastructures, deploy new VMs, and utilize Azure services.

VMware vSphere clusters are constructed atop hyper-converged, bare-metal hardware that shares nothing. The AVS cluster design is dedicated and isolated, meaning that no tenant's networking, storage, or computation is shared. Microsoft maintains VMware vSphere clusters in Azure to simultaneously meet performance, availability, security, and compliance requirements while providing unified management, networking, and operational controls.

Since AVS operates hybrid workloads across on-premises vSphere and Azure private cloud, it provides a single pane of glass for gradually integrating essential governance and operational management controls as the optimal execution method.

AVS roadmap

AVS is the quickest and most cost-effective method to quickly move and operate VMware in the cloud, thanks to its unique Azure hybrid features and ESU for Windows Server and SQL Server. AVS provides parity with on-premises infrastructures, allowing data center migrations to be hastened and cloud advantages to be achieved sooner. This symmetry also allows IT teams to use the same VMware skills, procedures, and investments established in on-premises VMware systems. AVS provides a straightforward method for extending and migrating existing VMware private clouds to operate natively in Azure.

Since the launch of AVS 2 years ago, Microsoft has continued to make enhancements to the solution while ensuring that global customers have access to the solution. As of the writing of this book, AVS is now available in 24 regions globally.

The following are the new features and enhancements that have been added to AVS as of January 2022:

- **VMware vSphere 7.0**: Microsoft upgraded AVS from vSphere 6.7, which is now globally available. Any new AVS deployment will be on the new vSphere v7.0.

- **Support for Public IP to the NSX Edge for AVS**: The availability of a new Public IP feature is now on AVS. Most client applications operating on AVS need internet connectivity. These applications require both inbound and outbound internet access. AVS Public IP is a streamlined and scalable solution for operating these applications. With this capacity, we can do the following:

  - AVS has direct inbound and outbound internet connectivity to the NSX-T Edge

  - The capability to receive at least 1,000 Public IPs

  - Protection against DDoS attacks against network traffic entering and leaving the internet

  - Enable support for VMware HCX (VMware VM migration tool) over the public internet

  - Global expansion to 24 regions

- Azure NetApp Files data stores for AVS

- JetStream DR for AVS

- VMware vRealize Log Insight Cloud for AVS is now generally available

- **VMware HCX Enterprise Edition**: VMware HCX Advanced is the default version that is deployed with AVS. Customers will need to submit a support ticket from the Azure portal to get an upgrade to HCX Enterprise. HCX Enterprise is available to customers at no additional cost.

You can also reference any new features added in 2022 by reviewing the *What's new in Azure VMware Solution* Microsoft doc (`https://learn.microsoft.com/en-us/azure/azure-vmware/azure-vmware-solution-platform-updates`).

Microsoft will continue to add more features and upgrades to AVS.

Best practices for planning, deploying, and managing AVS

As a private cloud on Azure, AVS is a specialized infrastructure bundle with VMware vSphere clusters. Three ESXi hosts are required for the initial deployment; however, you can add additional hosts one at a time, for up to 16 hosts per cluster. vCenter Server, vSAN, ESXi, and NSX-T Data Center are components deployed as a part of the solution. Workloads from your on-premises VMware vSphere systems may be migrated or expanded to AVS. On-premises resources, resources in a private cloud, and additional resources in the Azure public cloud may all be added to your AVS private cloud.

Many business and technical considerations must be made while planning your AVS adoption journey, including scoping, architectural design, assessment, implementation, and management. As with any successful project, setting your goals and success criteria early on is critical to building the proper solution for the company's requirements.

The following is a list of some of the best practices to follow when planning to adopt AVS:

- **On-premises environment assessment**: An assessment of your on-premises data center must be done to understand your workload inventory, what should be migrated, and help to create a migration timeline.

  You can use Movere, Azure Migrate, or do an RVTools export to get the needed data. The assessment will also help to determine the size of your AVS environment. An assessment will also help you decide your monthly cost for running your AVS environment.

- **Use case for adopting AVS**: Identify the use case for adopting AVS. There are a number of different use cases why a customer will decide to utilize AVS. Some of those use cases are as follows:

  - **Retiring a data center**: Some customers a looking to move out of their on-premises data center for various reasons, such as an increase in the overhead management cost of this data center

- **Need for an additional data center**: Customers who want to add another data center but do not want a traditional brick-and-mortar building

- **Customers looking to get migrate to the cloud without modernizing or re-architecting their applications**: AVS has proven to be one of the fastest and most simplified ways to get started in Azure without any modification to your existing applications and servers

Once the assessment is completed and the use case is identified, you should now focus your attention on the following business criteria:

- Timeline for migration to begin and be completed
- Identify the set of applications that will be migrated first (many customers do a proof of concept before migrating production applications)
- Understand the SLA for AVS
- Azure region
- Azure subscription
- How many AVS nodes are required

Architectural design for AVS

Following the assessment and use case identification, you will be able to build an architectural design in collaboration with Microsoft and a partner. Typically, as part of the discovery phase and technical review, the following subjects will need to be addressed and reviewed:

- **AVS overview**
  - AVS node size and quantity
  - Storage
  - Networking and connectivity
  - Access and identity
  - Management and monitoring
  - Compliance
  - Identity
  - Governance

- **Security and networking**

  - Azure landing zone design and integration

  - Firewall

  - ExpressRoute and WAN connectivity

  - Internet egress and ingress

  - L2 extension

  - Overall network flow

- **Migration and management**

  - Migration timeline

  - Business continuity and disaster recovery options

  - Backup tools and available options

  - Monitoring toolsets

  - Identity integration

After the preceding criteria have been established and agreed upon by the different domain owners (networking, security, and operations), you can begin the AVS adoption with a pilot or proof of concept. We advise you to deploy and test the following:

- **Pilot deployment:**

  - Deploy the AVS nodes

  - Configure network connectivity

  - ExpressRoute Global Reach to on-premises

  - Azure Gateway connection to Azure

  - HCX configuration

  - Configure any additional Azure native services and third-party applications

- **AVS pilot phase**: At the end of an AVS pilot, you will want to make sure that the success criteria that were set have been met by comparing the pilot to your established business and technical baselines.

- **AVS production environment**: Plan for the transition from a pilot environment to a production AVS environment. Implement a migration plan and understand your node capacity requirements for your production AVS environment.

Customers often want to optimize their existing applications in AVS. This is made possible through resource management, monitoring and security, storage, and other Azure services. Once your AVS environment is configured and operational, the hub and spoke network connection provides a smooth expansion path to other Azure native services.

Summary

In this chapter, we did a quick recap of AVS, its common use cases, and new additions to the solutions since the beginning of 2022; we also walked you through some of the best practices for planning, implementing, and managing your AVS environment.

AVS is the fastest and cheapest path to start your cloud migration journey. For customers who are considering adding a new data center for BCDR or want to migrate their on-premises VMware workloads without refactoring their infrastructure and/or applications, AVS is the perfect solution.

Make sure to have all of your domains (networking, security, and management) onboard during your planning and deployment phases. This will create the much-needed unified approach when getting ready for your cloud journey.

AVS is a VMware-verified solution. Microsoft continually validates and tests enhancements and upgrades to ensure the integrity of AVS and its platform. Microsoft manages and updates all hardware and software for the private cloud. In doing so, you can put your attention to where it belongs: on creating and executing workloads in your private clouds, where they can best serve your company.

Index

A

Active Directory (AD) 33

**Active Directory Domain
    Services (AD DS) 252**

alert rule

  configuring, for AVS 236-242

application layer 276

AVS deployment, prerequisites 63

  AVS cluster, deploying 66

  AVS resource provider, registering 64, 65

  basic information 66-69

  validation 69

AVS environment

  managing 231, 286

  monitoring 231

  platform monitoring and
      management 231-233

  VMware toolset recommendations 233, 234

  VM workload management
      recommendations 234, 235

AVS environment governance

  alerts for Service Health, planned
      maintenance, and security 259

  Azure-native services integration 259

  ESXi hosts access 258

  ESXi host usage 258

  financial governance 258

  governance failure-to-tolerate (FTT) 258

  governance, for host quota 258

  governance for network monitoring 259

  storage policy for VM template 259

  vSAN storage space 258

AVS host quota

  requesting 49

AVS management components

  /22 address space, requesting for 282, 283

AVS networking design

  recommendation 33

AVS resource name 48

  /22 CIDR IP segment, requesting 52, 53

  AVS workload network
      segments, defining 54

  host quota request 49-52

  host size 48

  number of hosts and clusters,
      determining 48, 49

  virtual network gateway, defining 55

**AVS's egress network traffic, through
    on-premises firewall 30**

  components, implementing 30

  consideration 31

AVS's ingress network traffic, through on-premises firewall 30
components, implementing 30
consideration 31
Azure Alerts
configuring, for AVS 235
Azure cloud ecosystem
challenges 24
Azure Files
advantages 261
use cases 261
Azure file share
creating 263-265
mapping, to AVS VM 266, 267
prerequisites 262, 263
Azure Firewall
adding, to vWAN 142-145
AVS traffic, routing to vWAN hub 149-151
policy, creating for AVS internet connection 146, 147
Azure infrastructure
AVS, connecting to 70-73
AVS connection, validating 73-76
Azure Monitor Metrics 236
Azure-native solutions integration 260, 261
Azure-native tools 250, 251
Azure Hybrid Use Benefits (AHUB) 251
integrating, with AVS 251, 252
native Azure integration 251
single point of support 251
unified licensing and consumption 251
unified VM management 251
Azure NetApp Files 185
Active Directory connections 185
Azure VMWare Solutions 185
capacity pool, creating for 188, 189
NetApp account, creating 186, 187
NFS volume, creating for 190-194

performance best practices 195-198
Share Protocol 185
subnet, delegating to 190
supported regions 195
Azure NetApp Files volume
attaching, to AVS cluster 194-197
creating, for AVS 185
prerequisites 185
Azure portal
used, for adding NSX-T segment 112, 113
used, for deploying HCX Advanced 82, 83
Azure Private DNS
reference link 33
Azure Privileged Identity Management (PIM)
Active Directory Domain Services (AD DS) 252
Azure region 47, 281
pairs 47, 281
Azure Site Recovery (ASR) 17
Azure subscription 46
flow 46
identification 45
Azure Virtual Network (VNet) 24
Azure VMware Solution (AVS) 3, 249, 269
alert rule, configuring 236-242
architectural design 289, 290
Azure Alerts, configuring 235
best practices 288, 289
business alignment 231
connecting, to on-premises environment 76
deployment, planning 279
enhancements 287, 288
enterprise-scale 8
governance, leveraging for 286, 287
hardware specification 270, 271
high-level architecture 6, 7, 272, 273
host quota request 282

Metric options 243, 244

network and connectivity topology 274

networking requirements 276

networking scenarios, with traffic inspection 277, 278

number of nodes, determining 281

overview 270

peering, with on-premises environment 77, 78

resource group identification 280

resource name 281

roadmap 287

Site Recovery Manager (SRM) 200

software specification 270-272

subscription identification 279

use cases, in enterprise 7, 273, 274

virtual network gateway, defining 284, 285

workload network segments, defining 283, 284

Azure Vnet 275

Azure vWAN 24, 131, 134, 275

advantages 133

architecture 24

AVS ExpressRoute circuit, connecting to hub gateway 139-141

creating 134

ExpressRoute Gateway, creating 136

features 131, 132

gateway size, modifying 141

prerequisites 134, 135

types 133

virtual hub, creating 136-139

B

BCDR support types

bidirectional protection 202

disaster recovery 202

planned migration 202

business continuity 16, 37

design consideration 17, 37, 38

C

capacity pool

creating, for Azure NetApp Files 188, 189

Center for Internet Security (CIS) 256

centralized identity management 253

CloudAdmin privileges for vCenter

reference link 35

compliance 19

corporate policy compliance 20

country or industry-specific regulatory compliance 20

data retention and residency requirements 20

Microsoft Defender, for Cloud monitoring 20

recommendations 256, 257

Workload VM backup compliance 20

compute profile 93

D

data-at-rest encryption 184

DDoS protection 254

destination network address translation (DNAT) 26

disaster recovery 16, 37

design consideration 17, 18, 39, 40

distributed denial of service (DDoS) 253

Distributed Firewall (DFW) 28

Domain Name System (DNS) 33

  configuring, for AVS 117

  forwarder, configuring 118-120

  name resolution, verifying 120-122

dynamic host configuration protocol (DHCP) 33

  configuring, for AVS 107

  prerequisites 107

  server, creating with Azure portal 107-112

E

egress from AVS with NSX-T 28

  components, implementing 28

  consideration 29

egress from AVS with NVA 28

  components, implementing 28

  consideration 29

End User License Agreement (EULA) 109

enterprise-scale, AVS 8

  pre-requisites 9

ExpressRoute authorization key

  creating, on on-premises ExpressRoute circuit 76

ExpressRoute circuit 139

  connecting, to hub gateway 139-141

ExpressRoute gateway 136

ExpressRoute Global Reach 269

Extended Security Updates (ESU) 8, 274

F

failures to tolerate (FTT) 175

fault tolerance 174

firewall policy 146

  associating, with hub 148

  rule, adding 147, 148

G

governance 19

  AVS environment governance 258, 259

  ESXi hosts access limit 19

  for workload applications and VMs 259, 260

  host quota 19

  recommendations 258

  storage space, on vSAN 19

H

HCX Advanced

  deploying, with Azure portal 82, 83

  deployment prerequisites 82

HCX appliance IP requirements 57

  port requirements 57

  reference link 57

HCX key 84

HCX Manager URL 84

HCX Mobility Optimized Networking (MON) 59

HCX Network Extension (NE) 58

high availability (HA) 37

high-end (HE) hosts 5

hub and spoke network topology 24, 275

I

identity and access management 16, 33

  Active Directory Domain Services (AD DS) 16

  Active Directory groups 16

  Active Directory sites and services 16

  least-privilege roles 16

  role-based access control (RBAC) 16

  vCenter privileges 34

  vSphere permissions 16

identity management for guest VM 253

Infrastructure-as-a-Service (IaaS) 38

Internet consideration design
options, AVS 153-155

managed SNAT service 154

public IP, to AVS NSX Edge 154, 155

J

jumpbox 73

L

L2 network extension

benefits 58

planning 59, 60

pre-requisites 58

recommendations 59

Layer 2 VPN (L2VPN) 58

Layer 4 (L4) 24

protocols 24

Layer 7 (L7) 25

protocols 25

M

managed SNAT service

features 154

management network 56

Metric options 243, 244

Microsoft Azure Backup Server
(MABS) 17, 37

Microsoft Enterprise Edge (MSEE) 4, 270

Microsoft vSAN Management
Storage Policy 174

migration types, HCX

Cloud to Cloud Migrations 81

cold migration 81

VMware HCX bulk migration 81

VMware HCX replication-
assisted vMotion 82

VMware HCX vMotion live migration 81

N

network and connectivity topology 10, 23

AVS's egress network traffic, through
on-premises firewall 30

AVS's ingress network traffic, through
on-premises firewall 30

challenges 10

network traffic flows 10-16

network connectivity 4

AVS hardware SKUs 5

AVS software specification 5

clusters 5

hosts 5

private cloud 5

Network File System (NFS) protocol 261

networking requisites, AVS 25

networking scenarios, AVS with
traffic inspection 26

by third-party NVAs, in hub vNet 31

egress from AVS with NSX-T 28

egress from AVS with NVA 28

secure vWAN hub with default route
propagation enabled 26

network mapping

creating 212-216

network profiles 91

network security 253

recommendations 253, 254

network security groups (NSGs) 190
network traffic inspection,
    by third-party NVAs in hub vNet 31
  components, implementing 31
  consideration 32
Network Virtual Appliance
    (NVA) 15, 24, 127, 275
NFS volume
  creating, for Azure NetApp Files 190-194
NSX-T Manager
  used, for adding NSX-T segment 114, 115
NSX-T segment
  adding, Azure portal used 112, 113
  adding, NSX-T Manager used 114-116
  verifying 116, 117
  VM, connecting to 122-125
NVA solution, for traffic inspection 156
  Azure Route Server, deploying 162, 163
  prerequisites 157
  Quagga NVA, deploying 163-168
  virtual network, creating 157-160

O

on-premises environment
  AVS, connecting to 76
  AVS connection, validating 78
  AVS, peering with 77, 78
on-premises ExpressRoute circuit
  ExpressRoute authorization key, creating 76
on-premises HCX Connector
  compute profiles, creating 93-98
  configuring 90
  network profiles, creating 91, 92
  Service Mesh, creating 98-101
  site pairing, adding 90
OWASP Core Rule Set compliance 254

P

public cloud solution
  management 20
  monitoring 20

Q

Quagga NVA
  learned routes, checking 170, 171

R

RAID-1 (Mirroring) FTT-1 174
recovery plan
  running 224, 225
  testing 223, 224
recovery point objective (RPO) 16, 201
recovery time objective (RTO) 16, 201
region pairs, in Azure
  advantages 47
  reference link 47
replication network 57
resource group 47, 280
  identification 47
return on investment (ROI) 258
reverse mappings 214
Route Server
  learned routes, checking 170, 171
Route Server peering
  configuring 168, 169

S

secure vWAN hub, with default route
    propagation enabled 26
  components, implementing 26
  consideration 27, 28

security 18, 252
  centralized identity management 18
  for identity 252, 253
  network security 253-255
  permanent access limits 18
  VM and guest application security 255, 256
Server Message Block (SMB) 185, 261
service-level agreement (SLA) 231
Service Mesh appliance status
  validating 102
SIEM system 254
site pairing 90
  configuring, in vCenter 206-209
Site Recovery Manager (SRM) 17
  BCDR support types 202
  deploying, in AVS 203
  in AVS 200
  installing, in primary and secondary
    AVS environments 203
  instances, connecting on protected
    and recovery sites 210
  mapping, creating 210-212
  protection groups, creating 220-222
  supported scenarios 201, 202
site-to-site (S2S) VPN 24, 275
software-defined data center
  (SDDC) 10, 23, 174
storage policy 174
  configuring 174
  default storage policy, specifying
    for AVS cluster 181-184
  listing 175-178
  prerequisites 175
  setting, for VM 178-181
subnet
  delegating, to Azure NetApp Files 190

T

traffic filtering 254
transparent data encryption (TDE) 255
transport layer 24, 275

U

unified firewall rule management 254
uplink network 57
use cases, AVS
  in enterprise 7
user-defined routes (UDRs) 190

V

vCenter
  site pairing configuration 206-209
vCenter privileges 34
  custom role, creating 35-37
virtual hub 136
  creating 136-139
virtual machine replication
  configuring 216-219
Virtual Network (vNet) 270
virtual WAN (vWAN) 131
VM
  connecting, to network segment 122-125
  deploying 122
  moving, to different network
    segment 125-127
VM and guest application security 255
  advanced threat detection 255
  code security 256
  database encryption and monitoring 255
  encryption, for guest VMs 255
  Extended Security Update (ESU) keys 256
  security analytics 256

vMotion network 57

VMStoragePolicy

setting 178-181

VMware HCX 56

activating 88, 89

features 56

migration types 81, 82

network segments, defining 56

prerequisites 87

required networks, determining 56

VMware HCX Connector OVA

deploying 86, 87

file, downloading 84, 85

prerequisites 84

VMware SRM 201

VMware syslogs, for AVS

diagnostic settings configuration 245-247

prerequisites 244

using 244

vRealize Activities (vROps) 259

vRealize Network Insight (vRNI) 259

vSphere Replication appliance

add-on, installing 205

components 204, 205

installing 204

vSphere Replication protection group

creating 220-222

vWAN security

Azure Firewall, adding 142-145

prerequisites 142

W

Web Application Firewall (WAF) 254

Wide-Area Network (WAN) 24

Packt.com

Subscribe to our online digital library for full access to over 7,000 books and videos, as well as industry leading tools to help you plan your personal development and advance your career. For more information, please visit our website.

Why subscribe?

- Spend less time learning and more time coding with practical eBooks and Videos from over 4,000 industry professionals

- Improve your learning with Skill Plans built especially for you

- Get a free eBook or video every month

- Fully searchable for easy access to vital information

- Copy and paste, print, and bookmark content

Did you know that Packt offers eBook versions of every book published, with PDF and ePub files available? You can upgrade to the eBook version at packt.com and as a print book customer, you are entitled to a discount on the eBook copy. Get in touch with us at customercare@packtpub.com for more details.

At www.packt.com, you can also read a collection of free technical articles, sign up for a range of free newsletters, and receive exclusive discounts and offers on Packt books and eBooks.

Other Books You May Enjoy

If you enjoyed this book, you may be interested in these other books by Packt:

Exam Ref AZ-104 Microsoft Azure Administrator Certification and Beyond - Second Edition

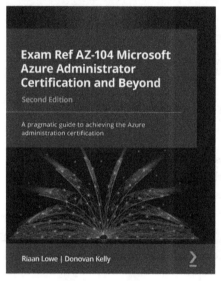

Riaan Lowe, Donovan Kelly

ISBN: 9781801819541

- Manage Azure Active Directory users and groups along with role-based access control (RBAC)
- Discover how to handle subscriptions and implement governance
- Implement and manage storage solutions
- Modify and deploy Azure Resource Manager templates
- Create and configure containers and Microsoft Azure app services
- Implement, manage, and secure virtual networks
- Find out how to monitor resources via Azure Monitor
- Implement backup and recovery solutions

Designing and Implementing Microsoft DevOps Solutions AZ-400 Exam Guide - Second Edition

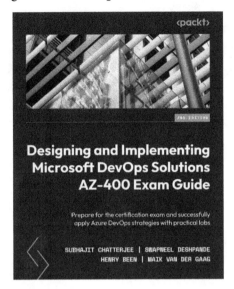

Subhajit Chatterjee, Swapneel Deshpande, Henry Been, Maik van der Gaag

ISBN: 9781803240664

- Get acquainted with Azure DevOps Services and DevOps practices
- Discover how to efficiently implement CI/CD processes
- Build and deploy a CI/CD pipeline with automated testing on Azure
- Integrate security and compliance in pipelines
- Understand and implement Azure Container Services
- Effectively close the loop from production back to development
- Apply continuous improvement strategies to deliver innovation at scale

Packt is searching for authors like you

If you're interested in becoming an author for Packt, please visit authors.packtpub.com and apply today. We have worked with thousands of developers and tech professionals, just like you, to help them share their insight with the global tech community. You can make a general application, apply for a specific hot topic that we are recruiting an author for, or submit your own idea.

Share Your Thoughts

Once you've read *The Ins and Outs of Azure VMWare Solution*, we'd love to hear your thoughts! Scan the QR code below to go straight to the Amazon review page for this book and share your feedback.

https://packt.link/r/1801814317

Your review is important to us and the tech community and will help us make sure we're delivering excellent quality content.

Download a free PDF copy of this book

Thanks for purchasing this book!

Do you like to read on the go but are unable to carry your print books everywhere?

Is your eBook purchase not compatible with the device of your choice?

Don't worry, now with every Packt book you get a DRM-free PDF version of that book at no cost.

Read anywhere, any place, on any device. Search, copy, and paste code from your favorite technical books directly into your application.

The perks don't stop there, you can get exclusive access to discounts, newsletters, and great free content in your inbox daily

Follow these simple steps to get the benefits:

1. Scan the QR code or visit the link below

https://packt.link/free-ebook/9781801814317

2. Submit your proof of purchase
3. That's it! We'll send your free PDF and other benefits to your email directly

www.ingramcontent.com/pod-product-compliance
Lightning Source LLC
Chambersburg PA
CBHW080928060326
40690CB00042B/3193